THE

SAGAN

CONSPIRACY

NASA'S UNTOLD PLOT TO SUPPRESS THE PEOPLE'S SCIENTIST'S THEORY OF ANCIENT ALIENS

DONALD L. ZYGUTIS

New Page BOOKS

This edition first published in 2017 by New Page Books, an imprint of
Red Wheel/Weiser, LLC
With offices at:
65 Parker Street, Suite 7
Newburyport, MA 01950
www.redwheelweiser.com
www.newpagebooks.com

ISBN: 978-1-63265-058-0

Library of Congress Cataloging-in-Publication Data
available upon request.

Cover design by Ian Shimkoviak/thebookdesigners
Interior by PerfecType, Nashville, TN

Printed in the United States of America
IBI

10 9 8 7 6 5 4 3 2 1

DEDICATION

In loving memory of my parents.

CONTENTS

INTRODUCTION

If we are poking around in neighboring worlds in our planetary system, then should not intelligent beings elsewhere in our solar system, as Lowell thought, or in other planetary systems, of which we now know there are many, shouldn't they be visiting here?

—Carl Sagan

Having studied his writings for more than 50 years, I consider myself somewhat of a Carl Sagan scholar—but I am hardly alone in that regard. Sagan, a Pulitzer Prize–winning author, had, and still has, millions of loyal followers all around the globe who enthusiastically devoured every word of every book he ever wrote. My bona fides is my claim to be the world's foremost authority on Sagan's ETI beliefs, particularly his work on ancient alienism. With this book I believe I am the first to recover and make public written material on ancient alienism that was crafted by Dr. Carl Sagan that has been suppressed and covered up by NASA.

It introduces something that few knew existed: an obscure 1962 scientific paper, researched and written by Carl Sagan with the funding of a NASA grant while he was a Miller Research Fellow at the University of California at Berkeley, then completed and published while he was working as an assistant professor of genetics at Stanford University—a document that reveals a comprehensive model of ancient alienism developed by Carl Sagan that adheres to all the rules and strictures of the scientific method. It is a model crafted by a scientist who expected it to be evaluated by his peers in the scientific community. It was peer reviewed and published in a respected scientific journal, *Planetary and Space Science,* an elite and specialized journal that is still in the business of publishing the research of the world's top space scientists—after which it effectively disappeared. Mysteriously, no one spoke of it, no one critiqued it, and no one referenced it. Through the years, as Sagan became famous, nothing was ever said of that paper. Today, it is probably fair to say that not one in a thousand people who think they know Carl Sagan are aware of its existence, and even fewer know that what he wrote in 1962 about interstellar spaceflight and past alien visitations to Earth, he continued to believe with great passion until his death in 1996 at the age of 62.

Thirty years later, after SETI (search for extraterrestrial intelligence) failed to detect an alien signal with radio telescopes, a mature Carl Sagan, by then the best-known scientist on the planet, was preparing to defy NASA, break his silence, and announce to the world his conviction that Earth is a visited planet. My charge is that in 1964, NASA officials, under the direction of the Pentagon, made the fateful decision to suppress the writings and muzzle the voice of a young Carl Sagan regarding his belief in ancient aliens.

This is an attempt to hold powerful institutions and individuals accountable for their actions. Of course I expect loud cries of indignation and categorical denials from representatives of all the accused parties, but, hopefully, along with their protestations and denials, they will answer the tough questions and address the straightforward issues that I raise in this book. If any of them want to confront me head-on, either in print or face-to-face in a public debate, I will be available.

The Conspiracy Theory

In this book I charge NASA with conspiring against Carl Sagan to keep knowledge of his belief in ancient aliens away from the public. Did Carl Sagan believe in ancient aliens? The paper written by Sagan, funded by a NASA grant, is indisputable proof that he did. Has the public been led to believe that Carl Sagan supported SETI's attempt to intercept an alien signal with radio telescopes? Absolutely. Was Carl Sagan convinced from the beginning that the odds that a radio telescope search would be successful were basically nil? Without a doubt.

There are two ways one can become a conspiracy theorist. The first is to concoct a sensational scheme out of thin air and then argue for it with all the fire and passion of a revival preacher, evidence be damned. The second way is to uncover a legitimate conspiracy, do your best to collect and assemble all the evidence you can, and then turn it into a book that reasonable people will find compelling—and then look for a publisher daring enough to put it into print. After all, it involves a man of tremendous scientific stature, Carl Sagan; the world's leading space agency, NASA; the world's largest ETI search organization, SETI; and, behind the scenes, lurking in the shadows, the Pentagon. We're not talking about small potatoes.

Other than to cite the agencies involved, I am not in a position to disclose the names and titles of individuals responsible for this travesty of justice, and for that I am thankful. My interest is not to indict people for wrongdoing as much as it is to focus on the positive and broadcast Sagan's ancient alien research to as many people as I can in as short a time as possible. With that goal in mind, I urge everyone to tell others about this book and pass it on to a friend when you are finished. Spread the word that Carl Sagan was an ancient alien theorist. The more people who know the truth, the sooner NASA, SETI, and others will be forced to admit to the truth.

The ultimate dream of any investigative reporter is to uncover a major conspiracy and expose it in print. But to be believable they have to produce facts, not innuendo. The evidence has to be reproducible and verifiable, not hearsay. Sagan's research paper is fact.[1] Sagan's ancient alien writings in *Intelligent Life in the Universe* is fact.[2] That the last paper he was working on before he died was about spacefaring extraterrestrials is fact.[3] Fill these facts in with NASA obfuscation and a veritable mountain of circumstantial evidence, and, in my estimation, there is more than enough material to prove a NASA conspiracy.

Like a book, the NASA conspiracy against Carl Sagan has three parts: a beginning, in 1964, when NASA suppressed his work on ancient alienism; a middle, from 1964 to 1990, when Sagan was not permitted to openly express his views on ancient alienism; and an ending, when he was preparing to defy NASA, break his silence, and take his ancient alien views public. Before he was able to complete his last paper, which was on ancient alienism, he died.

This book is based on the actual writings of Carl Sagan that conclusively prove that he believed in ancient aliens. Is there speculation? Of course. Do I at times extrapolate? I plead guilty. But one thing I don't do is fabricate. I am sharing what I know as an honest truth seeker and I stand prepared to be corrected if I am wrong. This is not the end of proving a conspiracy, it is only the beginning, and in short order I expect both a public and a scientific consensus to emerge that I am right.

Sagan looked at the subject of past alien visitations to Earth through the critical eyes of a professional scientist. With earned degrees from Harvard and the University of Chicago, and an advanced understanding of the scientific method, he anticipated that what he wrote about ancient aliens would be meticulously scrutinized and critically analyzed by his peers in the scientific community. Only those claims and calculations that managed to survive the harsh gauntlet of expert examination would be allowed to stand, and those only conditionally—because all scientific claims, regardless of how well founded they might be, are always open to challenge.

That was the way that Sagan wanted it to be. He never had any intention of circumventing scientific protocols and going straight to the public with his research. That high standard is reflected in the technical and scientific nature of the paper that he wrote while at Stanford University.

The volume that you are holding in your hands is, to my knowledge, the first book of what I expect to grow into a new literary genre: science-based ancient alienism. There are, today, some working scientists and skilled writers trained in science who believe it possible that Earth has indeed been visited by aliens. Unfortunately, their voices are not being heard because of a justifiable fear that if they were to openly state their views, they would be discriminated against by the scientific establishment (i.e., NASA). It is my hope that, before long, credible arguments and evidence of past alien visitations to Earth can be submitted in scientific papers by mainstream scientists, with the confidence that they and their work will be treated fairly and professionally by their peers. What can be successfully debunked will be summarily but respectfully abandoned, and what can't be debunked will be the subject of advanced investigation and analysis.

If my NASA conspiracy theory is true, how is it possible that the numerous authors who have written biographies about Carl Sagan over the years have failed to expose this? The fact is that several of them, and in particular Keay Davidson, have properly pointed out that Sagan believed in ancient aliens, and some, like him, alluded to this research, which I will refer to as the Stanford Paper.[4] But for a variety of reasons, not least of which is an almost-hallowed reverence for NASA among writers of popular science, no one had identified all the various pieces of incriminating evidence and assembled them in a way that results in the inevitable conclusion that NASA conspired against Carl Sagan. Eventually the truth was bound to come out. Better late than never.

The modern reading public has little tolerance for scripted narratives that scrupulously avoid reality. Most people, I suspect, would prefer to know the unvarnished truth about Carl Sagan. Sagan himself was a study in contrast and complexity: outwardly supporting an ETI search strategy involving radio telescopes that he didn't believe in, while inwardly clinging to an ancient alien theory that he knew NASA would never endorse. Now, it's time that the truth about Carl Sagan be laid bare as millions around the world pause to remember and honor a great scientist and an even greater human being on the 20-year anniversary of his death.

After reading this book, those who remain skeptical of my claim of a government conspiracy against Carl Sagan will be waiting for NASA and others to respond and prove me wrong. I have no problem with that. NASA deserves the right to defend itself. But if they have no rebuttal and are unable to refute my claims, will that convince the holdouts that my conspiracy theory is true? I would like to think so, but, at the same time, exposing a hallowed institution like NASA of a major cover-up against a celebrity scientist, who faithfully served for three decades as its face and voice to the world, is something so shocking that many will not be able to accept it, even in the face of overwhelming evidence.

For the past nine years I have been on an incredible journey, uncovering secrets about Carl Sagan and ancient aliens that a lot of powerful interests were hoping would never see the light of day. Though the primary purpose of this book is to blow the whistle on a NASA conspiracy, it's also to repair a reputation that will help make Carl Sagan better known and more relevant in the 21st century than he was in the 20th. Carl Sagan's legacy deserves to be that he was the scientist who gave the world hope for a brighter future, and that, if we continue to build on the progressive principles that the founders of human civilization were taught by their alien visitors, there is nothing that can deny us that destiny.

It takes teamwork to produce a book, and I would be remiss if I failed to mention some individuals who played key roles in the one you are reading: to my agent, Rita Rosenkranz; to my senior editor, Michael Pye; to associate editor, Lauren Manoy; and to my line editor, Jodi Brandon. If there were a literary equivalent of the Purple Heart, you would all be worthy recipients for enduring the foibles and follies of a fledgling author. And last, but certainly not least, to my partner, Sherry Subica, for her unflagging support and encouragement. My heartfelt thanks to you all.

CHAPTER 1

Carl Sagan, Raw and Uncensored

Over the course of what most people would call a spectacularly successful career, Carl Sagan became a trusted scientist to a generation largely raised by parents who were disillusioned by Nixon, Watergate, and the Vietnam War. For reasons entirely justified, our parents were suspicious of politicians, big government, and scientists who they saw as little more than pawns of the military-industrial complex. Then, seemingly out of nowhere, a young, cool, and very articulate astronomer began popping up on college campuses, the Johnny Carson show, and other places talking about the virtues of the scientific method, the best process ever invented for determining what is true and what isn't. He didn't speak down to us; he related to us—and he made sense. He made non-scientific people like me believe that we were capable of understanding what science was all about—that it wasn't the bogeyman that people like me were hearing from our religious leaders, or that young people were learning from their parents. Quite the opposite. Science, he insisted, was humankind's best hope for the long-term survival of our species.

Through advanced technology, Sagan assured us that we could clean up our polluted planet, control runaway population growth, achieve global peace, and bring nations and people together in common cause. We could explore space, visit other planets, and start colonies on the Moon and Mars. Carl Sagan, more than any other person on Earth, psychologically and intellectually prepared an entire generation to receive with open arms all the breathtaking possibilities of the modern computer age.

But people who think they know Carl Sagan invariably know him the way that influential individuals and powerful institutions in charge of his legacy *want* them to know him. All along, throughout the course of his 40-year professional career, Carl Sagan believed that advanced extraterrestrials exist and that they have been to Earth. Carl Sagan was an ancient alien theorist, convinced that human civilization was a gift from visiting aliens.

The truth is that from 1956, when Sagan was a 22-year-old whiz kid at the University of Chicago hobnobbing with Nobel laureates, until December 20, 1996, the day of his death, Sagan not only believed in ancient aliens, he single-handedly built a scientifically rigorous model that makes it possible for ancient alienism to hopefully, one day soon, become a legitimate field of inquiry.

In 1956, years before the SETI radio telescope experiment was launched, Carl Sagan saw clearly what NASA, SETI, and professional skeptics were either unable or unwilling to grasp: that in a 14-billion-year-old Universe and a 10-billion-year-old Milky Way Galaxy, if advanced aliens exist anywhere, they should have already discovered and visited the four-billion-year-old pale blue dot— Sagan's term—that we humans call Earth. A simple theorem developed by Italian physicist Enrico Fermi as early as 1943 remains to this day a powerful argument in favor of Sagan's suspicions. Fermi's Paradox notes the high probability of life evolving on many planets over vast stretches of time and space, and that many, many advanced extraterrestrials should exist, but so far, we have not seen any proof of or had contact with any such alien civilizations. The sheer number of statistically probable alien civilizations contrasted with the complete lack of any proof of any alien existence famously moved Fermi to ask, "Where are they?" Today, more than a half century after the Search for Extraterrestrial Intelligence began, we still don't know whether we are alone in the universe or if we have cosmic company. Sagan was optimistic:

> Studies of the origin of the solar system and of the origin of the first terrestrial organisms have suggested that life readily arises early in the history of favorably situated planets. The prospect occurs that life is a pervasive constituent of the universe. By terrestrial analogy it is not unreasonable to expect that, over astronomical timescales, intelligence and technical civilizations will evolve on many life-bearing planets.[1]

Sagan thought that in a few centuries, humans will have developed the technology for interstellar travel. If that is true, he pondered, shouldn't aliens, having civilizations possibly millions of years older and millions of year more advanced than ours, have already been to Earth? In the 10-year period between 1956 and 1966, he wasn't writing popular books, appearing on the Johnny Carson Show, or hawking the virtues of space exploration to the masses; he had his nose set to the grindstone, engaged in the most ambitious project of his life: to build an airtight science-based argument that Earth has been visited by advanced extraterrestrials.

Ancient Alien Theorist

If you watch ancient alien documentaries on television, you hear the phrase *ancient alien theorist* mentioned over and over. A theorist is someone who develops and espouses a theory, which the dictionary defines as:

Theory (def.)

1. The analysis of a set of facts in their relation to one another.
2. A belief, policy, or procedure proposed or followed as the basis of action.
3. An ideal or hypothetical set of facts, principles, or circumstances.
4. A plausible or scientifically acceptable general principle or principles offered to explain phenomenon.
5. A hypothesis assumed for the sake of argument or investigation.
6. An unproved assumption.
7. Abstract thought.[2]

A theorist, according to the scientific definition of the word, is a lot more than a guy with an opinion. They invariably hold multiple advanced degrees from reputable institutions, an extensive curriculum vitae, and they usually sit at or near the top of their chosen discipline. A good example is Stephen Hawking, the English theoretical astrophysicist who formerly held Isaac Newton's Lucasian Chair in the Royal Society, who has made significant contributions to the science of black holes. Contrast this with the ancient alien theorists of television documentaries, who never cite their academic credentials or professional accomplishments—because they don't exist.

Carl Sagan was an ancient alien theorist in the same way that Stephen Hawking is a theoretical astrophysicist. He held advanced degrees in multiple disciplines from institutions like Harvard and the University of Chicago. A complete list of his academic and scientific awards and achievements would require pages. Though his specialty was astronomy, he had an amazing cross-disciplinary education that qualified him to speak out on a broad range of subjects

relevant to space and to Earth. By any measure, Sagan had the perfect qualifications to be an ancient alien theorist—and that wasn't
by accident. He chose to major in astronomy and biology, areas that
would serve to enhance his credentials as an ancient alien theorist
and equip him with the knowledge and skills that would be needed
to find hard evidence to confirm his controversial hypothesis.

In 1962, years before Carl Sagan became the famous astronomer and extraterrestrial hunter that millions grew to know and
love, he held another title: visiting assistant professor of genetics at
Stanford University. Carl Sagan a geneticist? Yes indeed. Along with
an advanced degree in astronomy, Sagan held an earned degree in
biology from the University of Chicago.

The head of the genetics department—in fact, the man that
Stanford hired to established it—was Nobel Prize winner Joshua
Lederberg (1925–2008). Knowing that Sagan's Miller Research
Fellowship in astronomy at the University of California, Berkeley,
was about to end, Lederberg persuaded NASA to grant Sagan a
short-term appointment as assistant visiting professor in his new
school of genetics in Palo Alto, just across the bay. Lederberg had
a keen interest in the possibility of extraterrestrial existence, and,
from his willingness to allow Sagan to write the Stanford Paper in
his department and under his oversight, was clearly enamored of
the possibility that visiting extraterrestrials may have influenced the
human genome in a way that both qualitatively and quantitatively
separates humans from other animals. His lasting contribution to
modern astronomy is that he coined the term *exobiology*, which is
currently one of NASA's hottest new disciplines.

While he was at Berkeley, funded in part by NASA grant NSG-
126-61, Carl Sagan researched and wrote "Direct Contact Among
Galactic Civilizations by Relativistic Interstellar Spaceflight." In
1962, under Lederberg's oversight at Stanford, Sagan eventually
published the paper that laid out the scientific foundation for his
ancient alien theory. In the Abstract, Sagan clearly stated his claim

"that there is the statistical likelihood that Earth was visited by an advanced extraterrestrial civilization at least once during historical times."[3] In the Stanford Paper, Sagan arrives at the conclusion that "approximately 0.001 per cent of the stars in the sky will have a planet upon which an advanced civilization resides" and "statistics . . . suggest that the Earth has been visited by various galactic civilizations many times (possibly ~10^4) during geological time."[4]

In the late 1950s and 1960s there were scientists in the Soviet Union who interacted with Sagan on the possibility of past alien visitations, but he was the only legitimate ancient alien theorist in the West to develop a formal model that included a search strategy and then present it to his peers in the form of a scientific paper for their consideration. It was subsequently peer reviewed and published in a respected scientific journal, *Space and Planetary Science.* The extraordinary and completely unexpected pronouncement of past alien visitations to Earth, by one of the preeminent SETI theorists in the world, sent shockwaves throughout the astronomy community. Yet, in a vivid example of how Sagan was decades ahead of everyone else in the field, the Stanford Paper was a combination of existing SETI theory, hard science, and compelling logic. It is a model crafted by a scientist who expected it to be evaluated by his peers in the scientific community.

Perhaps it seems contradictory that Sagan was also an iconic skeptic who spent his entire life debunking UFO claims, dismissing then-current UFO sightings as unreliable[5], thus rejecting the modern UFO craze as pseudoscience. But ancient legends related to the early Sumerians, including some recorded in the Old Testament, led him to a stunningly different conclusion about the possibility of past alien visitations: Some of the gods of antiquity could have been, and more than likely were, visitors from outer space.

As Sagan narrowed the scope of his treatise, he refined his hypothesis by moving from a detailed explanation of how humans might someday achieve interstellar spaceflight, to positing that godlike aliens were physically on Earth, our Earth, establishing a human civilization through the Sumerians. Sagan's own ethic, now popularly known as the Sagan Rule, was that extraordinary claims require extraordinary evidence. His hope was for a discovery of direct evidence that would independently corroborate his hypothesis. His suggestion was that confirmatory empirical data might be found on the Moon, or possibly in ancient manuscripts. Man has since been to the Moon and they didn't find any evidence of alien activity. But what about ancient manuscripts?

The Land of Sumer

Most people erroneously believe that ancient alienism began in 1968 with the publication of Erich von Däniken's *Chariot of the Gods?*, which is why most people have the false impression that the subject is inherently pseudoscientific. But the truth is that 10 years earlier, in the late 1950s, Carl Sagan, working off a collaboration with Russian scientists and Nobel Prize–winning geneticist Joshua Lederberg, formulated a mature and science-based theoretical model that had extraterrestrials mastering interstellar spaceflight, exploring the Milky Way Galaxy, and visiting Earth at regular intervals.

In 1966, in *Intelligent Life in the Universe,* Sagan expanded on his controversial theory by suggesting that an alien signal of some sort might be found in ancient manuscripts related to the Sumerians, people of an unknown origin and with an unknown language who built the world's first high civilization; the Sumerians attributed their advancements to teachings from half-fish, half-human creatures, the Apkallu:

> Some years ago, I came upon a legend which more nearly
> fulfills some of our criteria for a genuine contact myth.
> It is of special interest because it relates to the origin of
> Sumerian civilization. Sumer was an early—perhaps the
> first—civilization in the contemporary sense on the planet

Earth. It was founded in the fourth millennium B.C. or earlier. We do not know where the Sumerians came from. Their language was strange; it had no cognates with any known Indo-European, Semitic, or other language, and is only because a later people, the Akkadians, compiled extensive Sumerian-Akkadian dictionaries.

The successors to the Sumerians and the Akkadians were the Babylonians, Assyrians, and Persians. Thus the Sumerian civilization is in many respects the ancestor of our own. I feel that if Sumerian civilization is depicted by the descendants of the Sumerians themselves to be of non-human origin, the relevant legends should be examined carefully. I do not claim that the following is necessarily an example of extraterrestrial contact, but it is the type of legend that deserves more careful study.

Taken at face value, the legend suggests that contact occurred between human beings and a non-human civilization of immense powers on the shores of the Persian Gulf, perhaps near the site of the ancient Sumerian city of Eridu, and in the fourth millennium B.C. or earlier. There are three different but cross-referenced accounts of the Apkallu dating from classical times. Each can be traced back to Berosus, a priest of Bel-Marduk, in the city of Babylon, at the time of Alexander the Great. Berosus, in turn, had access to cuneiform and pictographic records dating back several thousand years before his time.[6]

Berosus, a Greek historian and Chaldean priest, wrote three books on the history and culture of ancient Babylonia, including the legend of Oannes. The Oannes myth meets Sagan's criteria for a potential "contact myth," textual evidence that includes "a description of the morphology of an intelligent non-human, a clear account of astronomical realities for a primitive people, or a

transparent presentation of the purpose of the contact," indicators
Sagan believed should be investigated as possible alien intervention:

> At Babylon there was (in these times) a great resort of people
> of various nations, who inhabited Chaldæa, and lived in a
> lawless manner like the beasts of the field. In the first year
> there appeared, from that part of the Erythræan sea which
> borders upon Babylonia, an animal destitute of reason, by
> name Oannes, whose whole body (according to the account
> of Apollodorus) was that of a fish; that under the fish's head
> he had another head, with feet also below, similar to those
> of a man, subjoined to the fish's tail. His voice too, and lan-
> guage, was articulate and human; and a representation of
> him is preserved even to this day.
>
> This Being was accustomed to pass the day among men;
> but took no food at that season; and he gave them an insight
> into letters and sciences, and arts of every kind. He taught
> them to construct cities, to found temples, to compile laws,
> and explained to them the principles of geometrical knowl-
> edge. He made them distinguish the seeds of the earth, and
> showed them how to collect the fruits; in short, he instructed
> them in every thing which could tend to soften manners and
> humanize their lives. From that time, nothing material has
> been added by way of improvement to his instructions. And
> when the sun had set, this Being Oannes, retired again into
> the sea, and passed the night in the deep; for he was amphib-
> ious. After this there appeared other animals like Oannes, of
> which Berossus proposes to give an account when he comes
> to the history of the kings.[7]

The rise of an advanced civilization from primitive clans, who
for centuries had struggled for survival in mud-bogged villages, was
so meteoric that it was almost as though aliens had dropped out
of the sky and taught it to them whole cloth. Unlike the norm in

history, nothing was borrowed or stolen from neighboring civilizations, because there were no other civilized people.

Sumerologist Samuel Noel Kramer brings the contrast in sharp
relief. He writes of the Sumerians:

> In spite of the land's natural drawbacks, they turned Sumer
> into a veritable Garden of Eden and developed what was
> probably the first high civilization in the history of man.
>
> They devised such useful tools, skills, and techniques as
> the potter's wheel, the wagon wheel, the plow, the sailboat,
> the arch, the vault, the dome, casting in copper and bronze,
> riveting, brazing and soldering, sculpture in stone, engrav-
> ing, and inlay. They originated a system of writing on clay,
> which was borrowed and used all over the Near East for
> some two thousand years.
>
> Be he philosopher or teacher, historian or poet, lawyer or
> reformer, statesman or politician, architect or sculptor, it is
> likely that modern man will find his prototype and counter-
> part in ancient Sumer.[8]

The Sumerians were the first, and, in many respects, they were
the best. Empires that followed added almost nothing to what they
inherited from the Sumerians. What they invented and developed
remains, to this day, the vestigial contributions of a species gifted
with extraordinary intelligence and ambition—and, as Carl Sagan
was convinced, no small amount of help from visitors from outer
space. In a nutshell, the Sagan Model proposes that advanced
extraterrestrials, perhaps millions of years ago, mastered interstellar spaceflight. They likely visited our planet thousands of times
over the course of its evolution. Then, some 6,000 to 10,000 years
ago, they made their most dramatic move. While we humans were
still living like animals in small hunter-gatherer clans, godlike aliens
came to Earth and taught the Sumerians writing, architecture, agri-
culture, animal husbandry, law, mathematics, engineering, and basic

principles of moral conduct that individuals and families must have if they are to survive and thrive as city states and nation states. And, oh, by the way, they also taught the Sumerians how to make beer!

Sagan, never timid, was confident enough to ask hard questions that led to cryptic conclusions: If advanced aliens are god-like, and the gods of human religions are alien-like, what's the difference?

And: Could some of the gods of the ancients have been visiting aliens, and, if so, which ones, and where would confirmatory evidence most likely be found?

And: If alien interstellar spaceflight is scientifically impossible, as NASA and prevailing scientific thinkers insisted at that time, why does science call Fermi's Paradox a paradox? Why hasn't it been universally dismissed as just a really dumb statement?

These are scientifically legitimate questions that SETI, 50-plus years later, still refuses to grapple with in depth, even though they have the same intellectual and scientific gravitas as the Drake Equation, the foundational principle of modern SETI theory. Though almost everyone disagreed with Sagan, and many openly called him and his ideas kooky, his research was so thorough and his arguments so compelling that no one was confident enough to challenge the core arguments he developed in an academically approved manner. In fact, an official NASA publication from 1977 states: "We cannot rule out the possibility that we might stumble onto some evidence of extraterrestrial intelligence while engaged in traditional *archeological* or astronomical research, but we feel that the probability of this happening is extremely small [emphasis added]."[9]

Some of Sagan's peers within the scientific community, men like Frank Drake, rather than engaging the Stanford Paper head on, went behind Sagan's back by attacking it parenthetically. Without mentioning Sagan by name, they insisted that interstellar spaceflight that would carry aliens from their planet to ours, or some day in the future transport humans to where aliens live, was physically and forever impossible[10] because overwhelming technological challenges and Einstein's cosmic speed limit meant that aliens could not have physically traversed the vast distances of space to get from their planet to ours.

This consensus opinion led to the false impression that Sagan was proposing that aliens traveled from their planet to ours in about the span of a human lifetime, and that they did it in a nuts-and-bolts spacecraft that more or less resembled a 1960s NASA rocket ship. This was a complete fabrication and absolutely not what Carl Sagan believed. By pure innuendo, through scientific gossip, these screwy ideas were being falsely attributed to Sagan and then quietly promulgated throughout the astronomy community.

But let's face it: Anyone who believes in ancient aliens has to be able to explain how they got to Earth before speculating on what they did while they were here, and that means engaging in the interstellar spaceflight debate. Sagan understood this with absolute clarity, which is why the focus of his Stanford Paper addresses longevity, propulsion systems, velocities, distances, and colonizing strategies, not how aliens helped the Egyptians build the pyramids.

Think of an attorney representing ancient alienism in a court of law before a jury of skeptics. His first challenge would be to convince the jurors that it is scientifically possible for advanced extraterrestrials to have physically reached Earth. If he succeeded, he could then move on to step two, which would be to produce empirical evidence that they have been to Earth. Finally, if he got that far, step three would be to talk about what they did while they were here. Popular ancient alien theorists are making their case backward. They need to stop, take a deep breath, and go back to the beginning, where Sagan was. Only after we win step one can we advance to step two, and only after we win step two can we advance to step three.

A good step one argument is that if the human species, barely 400 years into the Scientific Age, is now contemplating going to the stars, how can anyone in good conscience deny the high probability that advanced extraterrestrials millions of years ahead of us, long ago mastered interstellar spaceflight and have already been to Earth? Plenty of reasons were given why humans couldn't travel to the stars, but no one wanted to address a key point in the Stanford Paper—that if, barely 400 years into the Age of Science, humans were even talking about it, what might aliens a million years more advanced than us be capable of?

Carl Sagan, the greatest alien hunter of all time, was convinced that spacefaring aliens have visited Earth on numerous occasions, but it is important to note that he didn't believe in the ancient aliens of popular culture—the ancient aliens prominently displayed on supermarket tabloids and on the History Channel. Like many other scientists and academics, Carl Sagan thought that finding direct and irrefutable evidence of advanced extraterrestrial existence would rank among the most important discoveries in human history. But as a trained scientist and a leader in the skeptic movement, Sagan spent a lifetime debunking the evidence cited by tabloid ancient alien leaders like Erich von Däniken and Zecharia Sitchin. It must be noted that Carl Sagan was as sympathetic as any trained scientist could possibly be to pop culture ancient alien advocates; his only grievance was that they violated scientific standards with impunity. Still, he never contested their underlying thesis: that aliens have been to Earth. It was an observation that he agreed with. But as a working scientist, he had no choice but to keep a respectful distance.

Carl Sagan believed in ancient aliens—but he attacked those who were publicly advocating for ancient alienism.[11] How can this be? Sagan was convinced that he was doing ancient alienism a favor by demanding that anyone who claims to be an authority on the subject abide by scientific principles that were established in the 17th century by men like Francis Bacon and Isaac Newton, who brought the Scientific Age into existence. He knew that setting a lower standard for ancient alienism than that demanded of other scientific claims would do long-term harm to the genre. Even worse, it would invite skeptics to associate ancient alienism with such patently pseudoscientific theories as young Earth creationism, Bigfoot, the Loch Ness monster, and unicorns. Is that the neighborhood that serious ancient alien theorists want to live in? Sagan didn't think so.

Sagan freely admitted that he had no hard evidence that would enable his ancient alien theory to be scientifically confirmed by independent testing and analysis. Without the original source materials in hand, this was uncertain testimony that the naysayers could dismiss as myth, which is what they have done. So what did Sagan do? He formally advanced his theory, advocated a search strategy, and then shut the hell up. Of course, he was hoping for a serious engagement with the scientific community that would have provided him the opportunity to defend his research, and that his search strategy would be implemented. But unlike most controversial scientific articles, the Stanford Paper never prompted a formal rebuttal from fellow scientists through peer-reviewed and published papers. By all scientific protocols, that should have happened, but it didn't.

The Stanford Paper was produced with the help of a NASA grant, which made NASA a partner with fiduciary interest. In 1963, after Sagan completed his document, it would have been normal and proper for NASA experts to study the paper to see what it got for its money. The absolute absence of any official follow-up analysis or response from NASA regarding the Stanford Paper is passive evidence of a concerted effort by NASA to distance itself from scientific research conducted with its commission and under its oversight by one of its own astronomers.

Though the Stanford Paper was subsequently peer reviewed and published, Sagan was roundly criticized by fellow astronomers. Frank Drake called it "bad science."[12] Others used even harsher invective. Sagan's controversial theory that extraterrestrials, the Apkallu, may have appeared to the Sumerians and taught them the ways of civilization was universally rejected on the grounds that interstellar spaceflight was impossible. Carl Sagan was personally vilified for even daring to broach the subject, and his subsequent denial of tenure at Harvard University, and, later, his rejected application for membership in the prestigious National Academy of Sciences (NAS) on two different occasions, were almost certainly due to his belief in ancient aliens.

An even harsher penalty was that the dissemination of his research at Stanford was so effectively suppressed that few people today know of its existence. For all intents and purposes, Sagan's Stanford Paper became as lost a manuscript as any that may be currently buried under the burning sands of Iraq and Iran. Today, it is probably fair to say that not one in a thousand people who think they know Carl Sagan's work are aware of its existence, and even fewer know that what he wrote in 1962 about interstellar spaceflight and past alien visitations to Earth, he continued to believe with great passion until his death in 1996 at the age of 62. The historical record is absolutely clear and unequivocal: Carl Sagan was a highly skilled ancient alien theorist. So how is it humanly possible for such an important fact about such a famous person to have been kept a secret for such a long period of time? No one spoke of it, no one critiqued it, and no one referenced it. Through the years, as Sagan became famous, nothing was ever said of that paper.

Why is NASA not recognizing Sagan's work?

The simple truth, I believe, is that the Pentagon, NASA, SETI, and professional skeptics have colluded in a conspiracy of silence to keep Carl Sagan's research and lifelong belief in ancient aliens away from the public eye. NASA, I am convinced, fully understood that Sagan's theories about ancient aliens were scientifically robust— technically and mathematically unassailable. My charge is that in 1964, NASA officials, under the direction of the Pentagon, made the fateful decision to suppress the writings and muzzle the voice of a young Carl Sagan regarding his belief in ancient aliens.

The motive behind this outrage was simple: For Carl Sagan, a NASA astronomer, to openly state that Earth, not space, may be the best place to look for hard evidence of extraterrestrial existence was a heresy that undermined NASA's very existence. Looking for signs of extraterrestrials in space was a NASA cash cow that brought it global attention. And NASA's space-orientated search for life in the Universe has immense value for the Pentagon and the military-industrial complex. Something had to be done to preserve NASA oversight of the search for extraterrestrial intelligence.

But one of NASA's most brilliant young astronomers concocted a scheme that, five to ten thousand years ago, had advanced extra-terrestrials on Earth, interacting with humans. Everyone knew that extraterrestrials, if they exist, live on distant worlds in other solar systems. By definition they are creatures of the cosmos, which was NASA's specialty. NASA is an agency that has its eyes to the sky, not on Earth. Its top employees, highly skilled astronomers, astro-physicists, and astrobiologists, have their attention fixed on space, not on terra firma. Sagan's strategy to look for evidence of extra-terrestrial intelligence was based here on Earth. If implemented, Sagan's Earth-based search strategy would reroute public interest and financial support in ETI away from the space sciences—and from NASA and the Pentagon—to disciplines like archeology and anthropology. At that time, NASA was still an embryonic organiza-tion with an uncertain future. In all probability, some individual, or, more likely, some group of individuals, inside the Pentagon felt they needed to disparage Sagan's Earth-based search strategy before it had a chance to gain traction and compete against it.

Not surprisingly, the SETI Institute, which features the Carl Sagan Center for the Study of Life in the Universe, also has shown no interest in spreading the news that Sagan was an ancient alien theorist. NASA and SETI are equally determined to keep the search for extraterrestrial intelligence from slipping away from them and into the domain of Earth-based sciences like anthropology, archeol-ogy, ethnology, and ancient manuscript analysis. Though Sagan died two decades ago, his fame still gives SETI a measure of credibility that it would not otherwise have. To have the name and reputation of Carl Sagan associated with ancient alien theory, which is where it properly belongs, rather than radio telescopes, would effectively pull the plug on what has been one of SETI's most effective promotional assets. That experiment began in 1960, and in the first few decades public enthusiasm was at a near fever pitch. Hollywood stars, business moguls, politicians—everyone, it seemed, was on the SETI bandwagon. For decades, NASA and its semi-autonomous subsidiary, SETI, relied on Sagan to be their face and voice to the world. With the charm and eloquence of a JFK and the raw intelligence of an Albert Einstein, Sagan defended the use of radio telescopes to intercept extraterrestrial intelligence (ETI) signals that NASA and SETI were convinced were being beamed our way. Has the public been led to believe that Carl Sagan supported SETI's attempt to intercept an alien signal with radio telescopes? Absolutely. Was Carl Sagan convinced from the beginning that the odds that a radio telescope search would be successful were basically nil? Without a doubt.

NASA scientists need to swallow their pride, revisit Sagan's research, and test any Earth-based empirical evidence that may confirm that Sagan was right. In short, the world needs more pure theoretical scientists like Carl Sagan, and fewer parochial scientists who clearly seem more intent on securing government funding for their pet projects than in finding the answer to the most important question any human can ask: Are we alone?

CHAPTER 2

SETI East vs. SETI West

Whenever possible, there must be independent confirmation of the facts.

—Carl Sagan, Baloney Detection Kit

In certain respects this book is about the long, complex, and, at times, uneasy relationship between SETI co-founders Frank Drake and Carl Sagan. Sagan passed away in 1996 of a rare disease, and Drake is nearing the end of an amazing life and career. The interactions between these two gifted astronomers and SETI pioneers, sometimes in concert and sometimes as competitors, cover the length and breadth of SETI history.

Despite the popular notion, backed by NASA and SETI, that the two men were in lockstep agreement on extraterrestrial theory, the truth is that they held stark and irreconcilable differences. Drake believed that interstellar spaceflight was nearly impossible[1], which, by definition, made theories of alien exploration and colonization of our galaxy, including past visits to Earth, wrongheaded. Carl Sagan believed just the opposite: that alien interstellar spaceflight and past alien visitations to Earth were the inevitable result of natural cosmic evolution. This is where Sagan showed his true depth of character. As a trained astronomer and an acknowledged expert in extraterrestrial theory, he could have, and was expected to, jump on the radio telescope bandwagon along with everyone else. But he didn't. In fact, in his writings Sagan openly voiced doubts that a radio telescope search would be successful. Sagan was convinced that the SETI strategy had too many conceptual flaws that were being intentionally overlooked or understated.[2] As a better option, Sagan agreed with Italian physicist Enrico Fermi: If aliens exist anywhere in the Milky Way Galaxy, they should have already been to Earth. Sagan thought that searching for evidence of alien visitations in ancient manuscripts held a lot more promise than listening for alien radio signals from space because "over large distances, starship communication will occur very nearly as rapidly as, and much more reliably than, communication by electromagnetic radiation."[3] That is not what NASA or Drake wanted to hear.

Drake vs. Sagan

In September 1959, Cornell physicists Giuseppe Cocconi and Philip Morrison co-authored a paper entitled "Searching for Interstellar Communication" that was published in the journal *Nature*. They were the first scientists to formally and publicly recommend that a search for an electronic alien signal be conducted with the recently invented radio telescope. In 1959, Frank Drake had begun implementing a search strategy using radio telescopes that was based on the assumption that alien interstellar spaceflight was impossible. The unexpected appearance of a peer-reviewed and published paper written by two prominent physicists scientifically validated what Drake was doing on the sly. The Morrison/Cocconi paper was a scientific windfall that had the effect of legitimizing his radio telescope search.

Unknown to Cocconi and Morrison at the time, a young radio astronomer, Frank Drake, was already preparing just such a telescope at the National Radio Astronomy Observatory at Green Bank

(West Virginia) for that purpose. He recalls his mindset in his 1992 book *Is Anyone Out There?*, which he co-authored with science reporter Dava Sobel:

> Suppose, I said to myself, that some alien civilization has a radio telescope just like ours. How close would they have to be to us for us to pick up their signals? I assumed their transmitters were no more powerful than our best ones. I guess I could have credited the aliens with far more advanced systems, emitting signals so strong that even the smallest telescope would hear them over interstellar distances, but that seemed like unfounded speculation to me.[4]

His groundbreaking endeavor, Project Ozma, wasn't successful in intercepting an alien signal, but it served to validate the Cocconi/Morrison paper by demonstrating the scientific viability of the process. Scanning space for alien signals with radio telescopes was here to stay—at least until it was proven a failure. It was about then that Carl Sagan, a young graduate student at the University of Chicago, first contacted Drake, inquiring about data that he had garnered with radio telescopes regarding the atmosphere of Venus. Though they differed greatly in appearance, style, and temperament, their common interest in extraterrestrial intelligence created a bond, sometimes strained, that was to last until Sagan's death.

Having already concluded that extraterrestrials have been to Earth and interacted with humans in historical times, Sagan witnessed firsthand the positive and dynamic impact the Morrison/Cocconi paper had on the Drake Model. If he hoped to compete, he knew that his model needed a published scientific paper as well.

In 1959 there were two nascent but operational SETI strategies, one that proposed using radio telescopes to search space, the other examining ancient manuscripts, in particular those related to the Sumerians. Though Sagan started his search before Frank Drake,

Drake was ahead because he had the backing of the Morrison/ Cocconi paper. Sagan had some catching up to do. He needed a formal scientific paper that would serve to legitimize his model, and he knew that he was the only one capable of writing it. In contrast to Drake, in 1959, all that Sagan had was the now-famous comment that Nobel Prize–winning physicist Enrico Fermi made years earlier: that if extraterrestrials exist anywhere in our galaxy, they should have already made it to Earth, so where are they?

Green Bank

Mathematics is the language of science, and Enrico Fermi was one of history's great mathematicians. Fermi's Paradox is, in essence, a mathematical construct that says given the age of the Universe (14 billion years), the age of the Milky Way Galaxy (10 billion years), and the age of Earth (4.5 billion years), it is inconceivable that aliens, if they exist, would not have been to Earth.

Drake had no idea that Carl Sagan had grasped the full significance of Fermi's simple but elegant logic that subtly deduced that any long-lived alien species, if they exist, would be so far ahead of us in science and technology that, of course, they would have already colonized the galaxy and been to Earth. How could they not? Implicit in Fermi's argument is that advanced aliens would, of course, have long ago developed the ability to travel to other stars. They wouldn't be very advanced if they hadn't. Considering the technological progress that humans have made in the brief time we have been in the Scientific Age, how would it be possible for long-lived alien civilizations millions of years into the Scientific Age *not* to have developed the ability to explore the Milky Way? Drake, aware of Fermi's comment, but not seeing any indication that anyone was doing anything with it, felt comfortable ignoring it.

In 1961, 11 men, then considered the foremost authorities in the world on extraterrestrial intelligence, gathered together in conference at Green Bank, West Virginia, to discuss how best to implement the power of radio telescopes in the search for an alien signal. Caught up in the rapture of knowing they were making history, they ostentatiously named themselves the Order of the Dolphin. Sagan was appointed secretary.

Drake knew that Sagan was interested in extraterrestrials and radio telescopes, but at that time he had no idea that the young astronomer, whom he describes in his book as "dark, brash, and brilliant," was engaged in a secret research project of his own that addressed the possibility that aliens have been to Earth. The same tight-lipped attitude that Drake had about Project Ozma, Sagan had about his suspicions that clues of past alien visitations to Earth might be located in ancient manuscripts. Had Drake known what Sagan was up to, it is doubtful that he would have invited him to the Green Bank Conference.

In preparing for Green Bank, Drake developed an equation that has become one of the most famous calculations in history:

$$N = R\ f_p\ N_e\ f_l\ f_i\ f_c\ L$$

N = The number of detectable civilizations
R = Rate of star formation
f_p = The fraction of stars that form planets
N_e = The number of planets hospitable to life
f_l = The fraction of those planets where life actually emerges
f_i = The fraction of planets where life evolves into intelligent beings
f_c = The fraction of planets with intelligent creatures capable of interstellar communication
L = The length of time that such a civilization remains detectable

This equation became the guiding outline for the conference, generating vigorous discussion on all the key considerations—except one: What percentage of advanced extraterrestrial civilizations exist long enough to have developed the ability to physically explore the galaxy? Enrico Fermi's observation and the intriguing possibility of alien interstellar spaceflight were never discussed. Drake has since admitted that this was not an oversight. It appears to me that the reason he left it out was because, if it had been included, it would have led to the inevitable conclusion that Fermi was right: that extraterrestrials could have and should have made it to Earth. Perhaps sensing the emergence of a competitive theory, Drake had good reason for ending the discussion of alien advancement with their invention of radio telescopes. To go hundreds or thousands or millions of years beyond that would have opened the door to alternative search strategies that he was determined to abort before they were conceived. But by then it was too late. The Sagan Model of ancient alienism had already been hatched.

The group at Green Bank estimated that the value of fc, the fraction of planets with aliens advanced enough to have developed radio telescopes, was 10 to 20 percent. These would be the ones that were approximately as advanced as we are, barely into the Scientific Age, and it was agreed that they were more than likely a minority. This left the majority of existing alien civilizations thousands or millions of years ahead of us in science and technology. What might they be capable of beyond sending radio signals? Incredibly, this commonsense question was never raised and, therefore, never discussed.

On Wednesday, November 1, 1961, Drake stood before the group and made an announcement: "Our best estimate is that there are somewhere between one thousand and one hundred million advanced extraterrestrial civilizations in the Milky Way."[5]

Sagan knew that even at the low end of this estimate there were more than enough long-lived advanced civilizations in our galaxy to almost guarantee that some of them would have developed the technology to physically come to Earth. It was as if the top SETI scientists in the world were serving up on a golden platter critical support for his unspoken thesis that Earth is a visited planet.

The Value of L

Both Drake and Sagan agreed that the calculations between N and L more or less canceled each other out. This resulted in a streamlined version of the Drake Equation: N, the number of detectable civilizations, equals L, the length of time that a civilization remains detectable.

All of the attendees at Green Bank agreed that there would be some percentage of alien civilizations that would be long-lived. The L factor distinguished those civilizations that destroy themselves within a thousand years after entering the age of science from those that manage to overcome self-annihilation. Sagan concluded that if that were true, some of those who became long-lived should have developed the ability for star travel. Of this group Sagan assumes "starship communication" would be the preferred method rather than electromagnetic signals.[6]

"Starship communication" is Sagan-speak for alien visitations to Earth. In effect, he was saying that searching here for an alien signal in ancient manuscripts was a more hopeful and reliable strategy than searching for an electromagnetic alien signal with radio telescopes. This was a subtle way of criticizing the Drake Model while at the same time trumpeting his own strategy.

Carl Sagan wasn't doing anything wrong in critiquing the Drake model. Scientists advancing innovative new theories routinely ask their peers to deconstruct and forensically dismember their ideas and kill them if they can. It is a solemn responsibility that all scientists not let any theory advanced by a peer have a free ride. Regardless of what the theory is or who the advocate happens to be, they are all required to pay their dues by being subjected to a brutal and relentless gauntlet of meticulous scrutiny so that only the strong survive, and precious time and resources are not wasted on the weak. In a blatant violation of this fundamental credo, the Drake Model was given a free ride, and the Sagan Model was rejected without being granted due process.

As a matter of conscience, Sagan could not go along with Frank Drake's thinking about interstellar spaceflight. It lacked imagination and was too narrow and restrictive. It failed to credit alien species thousands of years older than ours with abilities that we are likely to have if we survive that long. In a historic breach of scientific etiquette, there were those who were artificially insulating the Drake Model from critical scrutiny, while at the same time panning the Sagan Model in the most pedestrian manner imaginable. Sagan wasn't some wild-eyed, off-the-grid iconoclast. He was well schooled and deeply committed to the prevailing culture and standard protocols of his discipline. The evidence proves that Carl's critics are the ones who stand guilty of grossly abusing the scientific method.

In his Stanford Paper, Carl Sagan made the entirely plausible assumption that the odds that Earth has been visited by long-lived extraterrestrials in historical times are the same as the odds that a long-lived human civilization would visit other habitable planets in other star systems. Frank Drake held a position that was the polar opposite.

Sagan openly bashed the Drake Model in his writings, and he expected his model to be openly bashed in return. He anticipated and had the right to expect that blow and counter-blow would be struck in accordance with the scientific tradition of vigorous competition conducted in a spirit of fair play, transparency, and mutual respect. What he got instead was a secret kangaroo court verdict against his theory, and a coddling uncritical embrace of the Drake Model.

Of the Green Bank Conference, Drake wrote: "What bound us all together was our strange, strong conviction that the universe was widely populated. Among this group of people—and nowhere else I knew of—one could abandon oneself to the vociferous discussion of extraterrestrial life, without any hesitation, or embarrassment, or fear of ridicule."[7]

With all due respect to Frank Drake, I'm not sure how much "abandonment" there was at that historic gathering. Clearly, one man, Carl Sagan, was holding back. By that time, Sagan was already working on the high probability that advanced extraterrestrials would have developed the ability for interstellar spaceflight and, in all likelihood, have already been to Earth. This was all in Sagan's head, but he obviously wasn't comfortable bringing it up. Knowing that the theme of the conference was about radio telescopes, and thinking that he was going to be Frank Drake's future competitor, it was neither the time nor the place to throw down the gauntlet. Maybe he didn't want to be a party pooper.

Not only did Sagan's ancient alien theory raise serious questions about the wisdom of trying to establish contact with aliens using radio telescopes, he may have felt that the Green Bank participants were in no mood to either appreciate or critically evaluate the considerable amount of technical research and historical analysis that he had invested in his project, calculations that were soon to appear in his Stanford Paper and then greatly expanded in *Intelligent Life in the Universe*.

The Sagan Model was at a further disadvantage because it couldn't compete against the glamour of a new technology. When Sagan proposed his ancient alien theory, NASA scientists were already jumping on the radio telescope bandwagon. By any measure, astronomers thought that eavesdropping on the stars with a hot new invention that was specific to their discipline was a lot sexier than searching for a literary signal in dusty old manuscripts.

But, without question, Sagan appreciated the Drake Equation because it generated critical scientific support for his theory of past alien visitations. In fact, much of his Stanford Paper is an analysis and reaffirmation of Drake's mathematical model. Ironically, the Drake Equation, a key component in the scientific justification for SETI's radio telescope search, was equally applicable to the Sagan Model. It elevated Sagan's ancient alien theory from tenuous speculation to scientific credibility. It is highly doubtful that the Stanford Paper would have survived a peer-review process without Frank

Drake's contribution. In effect, Carl Sagan stole the secret energy source of the Drake Model right from under Frank Drake's nose and made it an integral part of his own model.

The Drake Equation was a robust, science-based argument for the existence of extraterrestrial intelligence. It became an effective tool that Drake and others used to persuade skeptics in science and government to open the purse strings and support a formal systematic search for alien existence. It convinced NASA to dedicate more radio telescope time to SETI to listen for an alien signal. Even the ultimate goal of having SETI officially recognized and supported with public funding by being written into the NASA budget was realized, until a penny-pinching Congress decided it was a luxury that the American taxpayer couldn't afford.[8]

The Launch of SETI

After the Green Bank Conference, momentum for a radio telescope search began to accelerate. Radio astronomers like Frank Drake knew that there was no limit to its potential. As radio telescopes grew more powerful and sensitive, almost every planet in every star system in the observable Universe would eventually be within reach, the only exception being the planet we live on.

Despite occasional setbacks, SETI continued to grow, and all kinds of scholars and scientists within the context of their respective disciplines began thinking about what an ETI might be like. Some wondered about their physiology, others about their art and music, still others about their technology. Drake and other radio astronomers garnered massive support by assuring everyone that an intercepted alien signal would likely answer all of their questions. Contact might even eradicate human ignorance and superstition and usher in an unprecedented Golden Age of Enlightenment.

Sagan, however, harbored serious doubts that extraterrestrials were sending electronic signals our way and had expressed his misgivings about radio telescopes in his Stanford Paper. As an alternative, Sagan knew that there was enough circumstantial historical evidence to justify an Earth-based search for evidence of past alien visitations. He also knew that NASA and his peers in the astronomy community would have absolutely no interest in it, and neither would archeologists, anthropologists, and other types of scholars who specialize in Earth-based research.

Sagan's logic was simple: If we humans are within a few centuries of becoming star travelers, advanced aliens "aeons" [dictionary definition: an indefinite period of time] more advanced than us should have been star traveling for a very long time, surely long enough to have reached Earth. But have they been to our planet? Sagan saw compelling evidence in ancient manuscripts that they have and argued for further research that related specifically to the Sumerians, the people who developed the world's first high civilization. Was the Sumerian Enlightenment a fluke of history, or, as Sagan envisioned, did it happen because primitive humans got a giant assist from advanced extraterrestrials who came to Earth from a distant planet in another solar system?

SETI scientists didn't believe that extraterrestrials could get to Earth, and archeologists and historians didn't want them here. By that time, Sagan knew that trying to get support and funding from NASA for an Earth-based search for evidence of past alien visitations would be futile because his model didn't have military assets. But he did what he could do. He stated his claim, presented his evidence and arguments, and was prepared to continue fighting for what he believed for as long as it took, which, in his mind, would be when a consensus developed that the Drake Model was a failure.

But the Pentagon and NASA had other plans for Sagan, and they had the leverage to make it happen. Perhaps they informed him in no uncertain terms that if he wanted to remain a NASA scientist, he had to abandon his ancient alien model and support the Drake Model. What choice did he have? If he had resisted, he would have summarily been thrown out of the astronomy community and publicly disgraced as a scientist. All his hopes and dreams would have been dashed; his long years of preparation would have been for naught. Sagan capitulated, but he never surrendered. His resolve was to live to fight another day.

Frank Drake came out of the Green Bank Conference the big winner, but with success comes the pressure to produce. The systematic search for an alien signal with radio telescopes became a reality, and the goal of its proponents was nothing less than to make the greatest scientific discovery in history, to forever alter the course of human history by proving that we are not alone—that other beings of high intelligence exist and are reaching out to us from distant planets. Despite SETI astronomers trying to downplay expectations, hundreds of millions of people around the globe, including leaders in science, government, business, and popular culture, enthusiastically jumped on the SETI bandwagon, hoping for and expecting a reasonably quick intercept.

The Victor and the Spoils

In *The Scientist as Rebel*, physicist Freeman Dyson writes about scientific revolutions precipitated by the invention of new tools, as espoused by Peter Galison, and those brought about by paradigm changing new ideas, an opinion embraced by Thomas Kuhn:

> The arguments between Kuhnian historians emphasizing ideas and the Galisonian historians emphasizing tools have continued to be vigorous. Historians trained in theoretical science tend to be Kuhnians, while those trained in experimental science tend to be Galisonians. Whether one chooses to emphasize ideas or tools is to some extent a matter of taste.[9]

With an engineering background, Frank Drake was a Galisonian who was enthralled with an exciting new tool: the radio telescope. He thought of a way to use it to possibly achieve the most exciting discovery in human history—to have it detect an incoming alien signal.

Carl Sagan was an idea man, a true Kuhnian. He began with the extraordinary possibility of there being intelligent life in the Universe beyond Earth, and then worked his way through all the permutations of what that might imply to arrive at the conclusion that, if long-lived alien civilizations exist, they would have already been to Earth.

In the early 1960s, the Galisonians prevailed over the Kuhnians. The tool of radio telescopes trumped the theoretical idea that aliens have been to Earth. Frank Drake won, Carl Sagan lost.

Seeing that he lost, Sagan threw his support behind the radio telescope experiment, knowing that the sooner it was proven not to work, the sooner the scientific establishment could turn its attention to his Kuhnian idea of past alien visitations. It had to have been discouraging for Sagan to see decade after decade go by and NASA and SETI stand behind Frank Drake, when it was completely obvious to him that the radio telescope experiment would never succeed. Despite its failure, NASA continued to throw its support behind the tool strategy, agreeing with Frank Drake that all that was needed was more time and more sophisticated and expensive tools.

The tool experiment has now been going on for more than 50 years without success. Isn't it time, in the interest of scientific and intellectual diversity, to let Carl Sagan and the Kuhnians have their moment in the sun, particularly because there may be evidence waiting to be tested that may confirm that aliens have been to Earth?

We know for a fact that Drake and Sagan both agreed that the Drake Equation was an effective tool that supported the concept of extraterrestrial existence, but how each man would go about using that tool was markedly different. Drake used it to argue for a radio telescope search for an electromagnetic signal from extraterrestrial civilizations who were too far away to be able to deliver their message in person. That effort was to become SETI, the Search for

Extra-Terrestrial Intelligence. Sagan's informed opinion was that the SETI search strategy was ill-conceived and misguided.

Still, to his great credit, Sagan was open to implementing the Drake Model—even enthusiastic—though in his mind the Drake Equation was proof that there are long-lived alien civilizations in our Galaxy that are sufficiently advanced to have developed the ability for interstellar spaceflight and physically come to Earth. He reasoned that if that was true, we should, along with a radio telescope search, be looking for evidence of past alien visitations in ancient manuscripts related to the people who built the world's first civilization, the Sumerians.

Sagan was more than eager to concede that advanced extraterrestrials had radio telescopes. It was kind of a no-brainer. He simply argued that radio communication was too inefficient for hyper-advanced extraterrestrials to bother with, particularly when they had the ability to physically travel to other star systems. Yet, despite his misgivings, Sagan invested great amounts of time and energy supporting the radio telescope search and, in fact, became SETI's global face and voice, and its most effective fundraiser.

Frank Drake, unfortunately, wasn't prepared to reciprocate. He was adamant in his refusal to accept that aliens—any aliens—would have mastered the technological capability for star travel. On the basis of that short-sighted conviction, he vigorously objected to any Earth-based search for evidence of past extraterrestrial visits, insisting that it was bad science and would be a complete waste of time. If Drake had been like Sagan, and admitted to at least a remote possibility of alien interstellar spaceflight, we can assume that he might have welcomed competition and supported the Sagan Model of ancient alienism, particularly because the cost of studying ancient manuscripts would have been miniscule compared to the hundreds of millions of dollars that have been poured into the radio telescope search. Drake's intransigence on this fundamental issue, and his stubborn unwillingness to allow competition, created a degree of separation between the two men that was to last through-out Sagan's lifetime.

What transpired at Green Bank informed Sagan that he was behind in the race, not to detect an alien signal, but to convince NASA leadership that his strategy, based in Fermi's logic, had more scientific merit than the Drake Model. He wasn't expecting NASA to abandon the Drake Model, but with a peer-reviewed and jour-naled scientific paper he had hopes that NASA would honor the scientific tradition of competing hypotheses and agree that both should be pursued and supported simultaneously.

Carl Sagan and Joshua Lederberg agreed with Enrico Fermi that if extraterrestrials exist, they should have already been to Earth, and in their minds there was ample evidence in ancient manu-scripts that they have been to Earth. To them it made sense to have another team of experts searching ancient manuscripts and exam-ining ancient artifacts to see if perhaps conclusive evidence of past alien visitations could be found here on Earth. But to justify forming a competitive search team they needed more than an offhand com-ment from a famous mathematician. They needed the equivalent of the Morrison/Cocconi paper that was published in *Nature*. They needed a formal scientific paper, and that is what Sagan produced in 1962 while he was at Stanford.

It soon became obvious that what Sagan and Lederberg had up their sleeves was a collaboration that was designed to split the nascent search for extraterrestrial intelligence into two competing camps. The East Coast team, based out of Cornell University in Ithaca, New York, had the 1959 Morrison/Cocconi Paper as its foun-dational document. Both Morrison and Cocconi were Cornell phys-icists, and SETI co-founder Frank Drake was a Cornell alumnus, so it would not be inaccurate to call those who believed in a space-based search for extraterrestrials the Cornell team, or SETI East. Based on what is admittedly circumstantial evidence, I'm convinced that the Stanford Paper was intended to be a charter document that would justify the formation of a team of scholars that would launch an Earth-based search for direct evidence of extraterrestrial existence. SETI West, the Stanford Team, presumably directed by Carl Sagan, would compete against SETI East, the Cornell team, under the direction of Frank Drake, in trying to be the first to make the greatest discovery in human history, to be the first to prove the existence of advanced extraterrestrial intelligence.

My contention that Joshua Lederberg and Carl Sagan made a bold end-run attempt to establish an Earth-based SETI West that was promptly shot down by NASA, would, if true, be a sensational news story. But is it true? Although I can't definitively prove it, every ounce of circumstantial evidence leads inevitably to that conclusion, so it's my story and I'm sticking with it. And the reason is simple: It is extremely unlikely that Sagan and Lederberg would suddenly come together at a major research university without there being a tie-in with the search for extraterrestrial existence. Furthermore, having an Earth-based SETI to complement and compete with a space-based SETI makes too much scientific sense for it not to be true. The fact that Carl Sagan was convinced that Earth has been visited by extraterrestrials makes it hard to arrive at any other conclusion.

Battling Nobel Laureates

Each of the two sides had a Nobel Prize winner to champion its respective position. Carl Sagan had Joshua Lederberg, and Frank Drake had Edward Purcell. Unfortunately for Sagan, Lederberg's support for the idea of past alien visitations to Earth, though it may have been vigorous in private, was muted in public. Purcell, in contrast, was quick on the attack. When first learning about the Sagan Model, he was highly vocal in his insistence that alien interstellar spaceflight, regardless of how technologically advanced any extraterrestrial civilization might be, was physically impossible[10], which meant that alien visitations to Earth were impossible, which meant that the Sagan Model was dead on arrival.

Sagan boldly and impiously called the elder Purcell's thinking wrong, arguing that long-lived alien civilizations would have worked out all the technological glitches and would have long ago embarked on interstellar travel, colonizing the galaxy along the way.[11]

Purcell could not claim to be unbiased. He had skin in the game. He won his Nobel Prize in physics by discovering the 21-centimeter line of hydrogen, which just happened to be the frequency that radio telescope aficionados thought that aliens would most likely use to transmit their signal on. Purcell had a personal interest in killing Sagan's ancient alien theory before it got off the ground. Had radio telescope SETI successfully intercepted an alien signal on the hydrogen frequency, Purcell would have been a primary benefactor. It would have added considerable luster to his fame and legacy.

Despite his compromised position, Purcell garnered critical support opposing the Sagan Model from other leading scientists, including Stanley Miller, one of Sagan's close friends and associates, who wrote of Sagan's theory: "It raises the question of whether he is a serious guy."[12] At the same time, a young Frank Drake, sensing an opportunity to secure his own career as a SETI radio astronomer, joined in on the pile-on, calling the Sagan Model "bad science."[13] Carl Sagan's stunningly simple and elegant theory never had a chance, and all the attacks against it were related in one way or another to the belief that interstellar spaceflight was impossible. Remove that one obstacle, and the Sagan Model becomes perfectly rational.

This bullying and name-calling against Sagan and his ancient alien theory made a mockery of the scientific method, which is designed to encourage competition. Did the peer harassment begin after the Pentagon issued specific orders to new NASA administrator James Webb that Sagan's theory needed to be terminated and abandoned because there was zero chance that it would serve any useful military purpose? In secret, I'm convinced that the Pentagon never bought into Purcell's argument that interstellar spaceflight was impossible, because, even at that early time, it was already developing long-range plans to colonize space. Through NASA, and with abject hypocrisy, it lent its support to Purcell and others who were blasting Sagan for suggesting that aliens engage in star travel.

Still, a two-team, two-model approach, if it had materialized, would have been compatible with the finest scientific tradition, when different teams of scientists, working off different sets of assumptions and methodologies, openly and knowingly compete against one another to be the first to make a major scientific break-through. Unfortunately, NASA leadership stepped into the fray, and unilaterally and without scientific justification decided that the Sagan Model had to go away.

A recent example of how effective competition in science can be is the remarkable story of how dark matter and dark energy were discovered. In the 1990s, an East Coast team of astronomers out of Harvard competed against a West Coast team of physicists at Berkeley to be the first to use supernova as a standard candle to measure the rate of expansion of the Universe, a dramatic, tension-filled race admirably chronicled by Richard Panek in The 4% Universe.[14] The result was the simultaneous discovery of two mysterious substances that comprise 96 percent of the Universe. Appropriately, both teams shared the 1998 Nobel Prize in physics.

Had this approach been adopted by NASA, the United States and the world would have been treated to a scientifically rich and interactive process that would have, early on, identified the strengths and weaknesses of both models and, as an added bonus, would have prevented the void that was created that was soon to be filled in by the questionable characters who peddle tabloid ancient alienism. The fact that reputable scientists broke with tradition and emotionally and irrationally rallied against allowing the Sagan Model to compete in a fair and open forum against the Drake Model suggests that behind the scenes there were dark forces at work.

Bad Science?

Sagan's research resulted in an essay that laid out the blueprint for an alternative strategy for searching for an alien signal that didn't involve radio telescopes. It was a direct challenge to SETI co-founder Frank Drake because it was based on the assumption that any advanced extraterrestrials worthy of that name would almost certainly engage in interstellar spaceflight, something that Drake and most other space scientists in that day didn't believe was possible. Sagan's Stanford Paper featured the stunning theory that extraterrestrials have physically been to Earth, and it included detailed information about where he thought a signal they left behind might be located.

In a brilliant example of inductive logic, Sagan blended all the key components of existent SETI theory into a single seamless proposition. Besides meticulously working his way through rocket propulsion systems and the Drake Equation, and factoring in Fermi's Paradox, he posited that advanced alien civilizations exist, have mastered interstellar spaceflight, and have been to Earth. He went on in his paper to reference some specific legends from antiquity that he believed supported his thesis, including accounts from the Old Testament that describe meetings between humans and godlike beings.

In developing an alternative search strategy, what Sagan and Lederberg failed to realize was that NASA was (and still is) a government bureaucracy first, and a scientific establishment second. Government bureaucracies are monopolies that are intrinsically opposed to competition. Just as your state department of motor vehicles has no interest in competing against other departments of motor vehicles for your patronage, in 1963 NASA administrator James Webb—a Washington insider, not a scientist—would have had little interest or motivation in allowing an Earth-based search strategy that had no military value, to compete against a space-based model that had military assets. As a result, the Stanford Paper was not just turned down, it was crushed and intentionally hidden from public view. Under no circumstances would there ever be a SETI West at Stanford University, and Carl Sagan would never be allowed to be the director of a competitive team that would challenge the Cornell team. In the end, science and humanity lost, and the military-industrial complex won.

In defense of James Webb, he was hired to do one job: fulfill President Kennedy's pledge to send American astronauts safely to the Moon and back by the end of the decade. Everything else was secondary. Perhaps in his mind, to succeed he had to have everyone affiliated with NASA, even if it wasn't directly related to the Moonshot, on the same page, pulling in the same direction. Carl Sagan's ancient alien theory had the misfortune of not fitting into the zeitgeist of the times, while the radio telescope search did. In the often-murky and conflicted world of situational ethics, Webb's decision to abandon and suppress the Sagan Model may have been corporately right because it didn't serve the interests of the Pentagon, but it was morally and scientifically wrong. As a result, Sagan's theory was not only unfairly savaged on non-scientific grounds, he was denied the due process that is ordinarily afforded professional scientists who play by the rules.

And what about the scientists at Cornell? Were they celebrating the fact that they wouldn't have to compete against Carl Sagan and a robust challenge from a Nobel Prize–winning scientist at a world class university? Were they delighting in the reality that they would be the only game in town, even though they knew that the Stanford Paper was an imposing document, and that not having a scientific Earth-based search would likely create a vacuum that would soon be filled by pseudoscientific charlatans?

You bet! Instead of being forced to play hardball against a worthy opponent, the Cornell Team could relax and play intermural softball. And Carl Sagan? After the NASA rejection of his paper, he left the West Coast and moved to the East Coast, first to become a professor at Harvard University, and then, after he was denied tenure under suspicious circumstances, to join the faculty at—you guessed it—Cornell University! The subjugation of a great scientist passionately committed to a bold new plan to make contact with ET was complete.

The reality, put bluntly, is that Carl Sagan got screwed, big time, and though there is no way to undue the personal anguish and humiliation he suffered in the vulgar way his model was rejected, his dignity can still be restored and the damage to his reputation mitigated by implementing the search strategy he first proposed in 1962 and that he planned on reintroducing in the 1990s. As a testimony to its enduring strength, the Sagan Model remains as hopeful and viable today as when it was first crafted.

Most of those in the Pentagon and at NASA who conspired to bury Sagan's ancient alien research are no longer with us. Here is hoping that the few who remain, including Frank Drake, will take the high road, step forward, and publicly confess that in 1964 there was a concerted plot by the Pentagon and NASA to stop Carl Sagan before he had a chance to implement his alternative search strategy. Later, in 1984, when SETI incorporated itself as a private entity, it established its headquarters at Mountain View, California, right next door to the Stanford campus. Presumably, it was so that it could be close to deep-pocketed Silicon Valley philanthropists who gifted it with millions of dollars, but that location also allowed it to keep tabs on any independent thinking scientists at Stanford University who may not have given up on the idea of challenging the radio telescope experiment.

One cannot possibly get more unscientific than to announce a result without benefit of an investigation, particularly when the announced result runs against the grain of scientific consensus. Yet this is what NASA and SETI have done in stating, carte blanche, that there is no evidence that extraterrestrials have been to Earth. An individual new to the discussion might assume that this strong definitive declaration has been made after an extensive search, when the truth is that there has been no search. The Earth-based search model proposed by Carl Sagan was never activated. Sagan's ancient alien research would remain in limbo for the next 50-plus years, and is only now being reintroduced and revisited.

At the time of the Green Bank Conference, host Frank Drake clearly didn't know about Carl Sagan's strong belief in interstellar travel, or about his conviction that aliens had visited Earth in historic times. In his book *Is Anyone Out There*, Drake writes about what he mistakenly believed was the unanimous opinion among the attendees that interstellar space travel was impossible:

> The possibility (of intercepting an electronic alien signal) was infinitely more plausible than the arrival of an alien spacecraft. There was absolute consensus on this point—that space was too vast to permit easy physical visitations between civilizations. Achieving speeds high enough to complete interstellar voyages in reasonable times would make energy demands that were too great, even for very advanced civilizations. "Contact" would be in the form of electromagnetic signals passing between worlds at the speed of light. (No matter how far into the realm of the fantastic our ruminations took us, we didn't even discuss the notion of interstellar travel at Green bank, on the grounds that it was irrelevant).[15]

This, in a nutshell, was Frank Drake's argument against interstellar spaceflight, and it is surely one of the most convoluted analyses in the history of science. With complete disregard for the major technological advances that long-lived alien civilizations would be expected to achieve, Drake superimposes the limitations of existing human technology on extraterrestrials who could easily be many millions of years older than us, and that many years ahead of us in science and technology.

Sagan's best estimate of the number of advanced civilizations in the Milky Way with technology beyond our own was between one and ten million, about in the middle range of what was determined at Green Bank by the Order of the Dolphin. He believed that most, if not all of them, would have developed the technology for interstellar spaceflight. His next consideration was how long it will take humans to develop that capability. Section 3 of the Stanford Paper, entitled "Feasibility of Interstellar Spaceflight," addressed the technical difficulties of velocity, fuel, and longevity:

> The purpose of this Section is to lend credence to the proposition that a combination of staged fusion boosters, large mass-ratios, ramjets working on the interstellar medium and trajectories through H II regions is capable of travel certainly to the nearest stars within a human shipboard lifetime, without appeal to as yet undiscovered principles. Especially allowing for a modicum of scientific and technological progress within the next few centuries, I believe that interstellar travel at relativistic velocities to the farthest reaches of our Galaxy is a feasible objective for humanity. *And if this is the case, other civilizations, aeons more advanced than ours, must today be plying the spaces between the stars.* [emphasis added][16]

Drake's insistence that there was "absolute consensus" among the Green Bank conferees that interstellar travel was impossible suggests that all 11 conferees, including Carl Sagan, discussed the subject at length. Isn't that what the word *consensus* means? Yet, in the same breath, Drake states that the subject of interstellar travel was never even mentioned on the grounds that it was irrelevant. It doesn't make sense.

Sagan, however, never apologized for or retracted his research. Quite the opposite: Working against tremendous institutional pressure from NASA to keep his heretical ideas of interstellar spaceflight and past alien visitations to himself, he brazenly expounded on his theory in a 1966 book he co-authored with Russian astrophysicist I.S. Shklovskii. In *Intelligent Life in the Universe*, he devotes two entire chapters to the subject. The response from the astronomy community was predictable. They redoubled their efforts to censure his research by prohibiting discussion of interstellar spaceflight and past alien visitations to Earth in SETI literature (except in the form of ridicule) and at official SETI conferences. In 1964, NASA officials were insisting that neither aliens nor humans would ever, could ever, physically travel from one star to another. The NASA moratorium on interstellar spaceflight became an unmovable object that countered the irresistible force of Fermi's Paradox that stated that if aliens exist anywhere in the galaxy, they should have been to Earth.

But the handwriting was on the wall. Seeing that there was no support for an Earth-based search strategy, Sagan went along with the radio telescope plan, knowing all the while that it was doomed to fail. Sagan would become the public image and voice for NASA, and SETI's most effective fundraiser, yet another reason why almost everyone erroneously assumes that he was completely convinced that a radio telescope search was the best way to find evidence of extraterrestrial intelligence. The truth, which neither NASA nor SETI is willing to admit, is that Sagan knew from the beginning that the SETI experiment was a dead man walking.

CHAPTER 3

SETI at Sunset

*Propriety feelings are of course
offended when a scientific hypothesis
is disproved, but such disproofs are
critical to the scientific enterprise.*

—Carl Sagan

I strongly suspect that the leadership team at the SETI Institute knows full well that the hypothesis that their alien search is based on has been thoroughly falsified, and that the time is fast approaching when the doors of its Mountain View, California, headquarters will close and the experiment will officially be shut down. The last of SETI's glory days was decades ago. In 1992, for example, Frank Drake, certain that the search model he invented was on the brink of success, wrote: "This discovery, which I fully expect to witness before the year 2000, will profoundly change the world."[1]

Much to Frank Drake's consternation and embarrassment, SETI's discovery of an alien signal wasn't as imminent as he thought. Almost 25 years after penning these boastful words, an alien radio signal still hasn't been intercepted, and the SETI experiment is widely regarded in the scientific community as a boondoggle. Unfortunately, Frank Drake has been so deeply invested in SETI and radio telescopes over his long career that he knows that, at his age, there is no way out of the corner he has painted himself into. His destiny, his legacy, will forever be attached to radio telescopes, and an alien radio signal that never came.

There is still time, however, to save the legacy of Carl Sagan from the same plight. Drake's caustic description of Sagan's Stanford Paper as "bad science"[2] is evidence that he knew about Sagan's belief in past alien visitations. Now, in his twilight years, it would be magnanimous of Frank Drake if he would come clean and tell the public the truth: that Carl Sagan was a life-long believer in ancient aliens and that he doesn't deserve to be associated with SETI and radio telescopes in any deep scientific sense.

This, however, is not likely to happen because it appears that Drake may be holding a grudge. He has never apologized for stating that Sagan was doing bad science when he wrote his Stanford Paper, and he evidently continues believes this so strongly that, even in light of the fact that his anti-interstellar spaceflight argument has been negated, he is still not willing to recognize the Sagan Model of ancient alienism as a scientifically legitimate ETI search strategy.

This raises a very uncomfortable question: Has Frank Drake, in his almost-fanatical belief that extraterrestrials will send us electronic signals, abandoned his training as a professional scientist to the point where he cannot find it within himself to even acknowledge the existence of the Sagan Model of ancient alienism, much less allow it the opportunity to compete in an open marketplace of ideas? Is he, in effect, playing God by, *a priori*, rejecting the Sagan Model without allowing new evidence that supports that model a chance to be tested? Could it be that Drake is still bothered that Sagan, while he was at the 1961 Green Banks Conference, was secretly putting the final touches on his ancient alien theory and that all he needed to make it fit for publication in a scientific journal was the equation that he—Frank Drake—developed?

In the 16th century, another Drake—Sir Francis by name— left the comforts of his native England and sailed around the New

World. The name of his flagship was the *Revenge*. Could it be that Frank Drake is the captain of a ship of a different kind that bears the same appellation? I believe that for the past 50-plus years, Frank Drake has been one of the men most responsible for keeping Sagan's name and reputation attached to radio telescopes.

In 1992, a cocky Frank Drake had the champagne on ice and the party balloons ready to be released. Throwing scientific caution to the wind, he promised the world that SETI radio telescopes would detect an alien signal by the year 2000. In contrast, in 1995, Carl Sagan, showing extreme caution, wrote the following about the chances of his ancient alien model being proven: "I didn't imagine that this would be easy or probable, and I certainly did not suggest that, on so important a matter, anything short of iron-clad evidence would be worth considering."[3]

Now, 20 years later, science writer Lee Billings, after an interview with Drake, wrote: "Drake sighed. 'These days I think that more-advanced technical civilizations will probably prove more difficult to detect than younger ones,' he said."[4]

Today, the Drake Model and the Sagan Model are on opposite sides of the horizon. The Drake Model, with its radio telescopes falling into disrepair, is at its sunset, close to disappearing entirely from the scientific line of sight. In contrast, the Sagan Model is rising, growing brighter with each passing year. There is no better indicator of what is happening than Drake's current position on interstellar spaceflight, which hasn't changed since 1960. He told Billings: "Of all the things we might someday do, I don't think we'll ever colonize the stars."[5] "Ever" is a long time, conceivably lasting for millions or even billions of years. Now Frank Drake is likely the only person in the world who still clings to the depressing and outdated notion that interstellar spaceflight is impossible, and that the human dream and aspiration to someday travel to other planets in other solar systems is nothing but an empty fantasy. More than any other single individual, I believe that Frank Drake is the man most responsible from keeping the Sagan Model, which is premised on interstellar spaceflight, from being revisited. Frank Drake has a ball and chain attached to Sagan's legacy, and he appears in no mood to set him free.

A Tale of Two Models

Frank Drake makes it clear in his writings that the central justification for using radio telescopes to search for an alien signal is because he is cocksure that alien interstellar spaceflight is impossible. Because aliens can't reach Earth physically, he argues, they have no choice but to communicate over the vast distances of space electronically. But if we accept Fermi's Paradox, Drake's rationale for his radio telescope experiment breaks down in the face of simple arithmetic and irrefutable logic.

Add to this the indisputable fact that after more than a half century of using radio telescopes, SETI hasn't detected anything, which is entirely consistent with Carl Sagan's argument that, having the choice, an advanced alien civilization would choose interstellar spaceflight over sending radio signals. As Sagan pointed out in his Stanford Paper, a personal visit by advanced aliens to an emergent species would be infinitely superior to transmitting electronic information that would take thousands of years to get to where they wanted it to go, and, even if intercepted, would likely be unintelligible, and, even if decrypted, would require thousands of additional years before the reception of a response.[6]

One has to ask: Why was the scientific establishment so hard on Carl Sagan and so soft on Frank Drake? Why are professional skeptics not attacking the Drake Model when it is stunningly obvious that it has failed? One reason may be that everyone sees Frank Drake as a humble type of guy, whereas Carl Sagan, with his supreme confidence and keen intellect, often came across as arrogant. Even if that were true, which it isn't, personal likeability or un-likeability should never be factors in determining who wins and who loses in a scientific competition.

Incredibly, professional skeptics and much of the public still act like the radio telescope experiment is adorned in the finest fashions, but the truth is that the king is naked. The continued implementation of the Drake Model is more than an exercise in futility; it's beginning to bear an uncomfortably close resemblance to pseudoscience. Over its 50-plus years, it has produced no evidence and there is no longer a scientific justification for its existence. Yet it continues.

Fortunately, there is an easy way that Frank Drake can dispel any suspicion that he is preventing the Sagan Model from being revisited out of revenge. All he has to do is to openly admit that Carl Sagan was an ancient alien theorist and acknowledge that the Stanford Paper has scientific validity.

The Drake Model is on life support while the Sagan Model, having been rediscovered, has gotten a new lease on life. Now it's time to lay both on the table of critical scientific scrutiny and see which fares the best. But for that to happen, Frank Drake needs to sign off on the project, and he gives no indication that he is willing to cooperate. He is still making the excuse that there is no possibility of interstellar spaceflight, even for advanced aliens millions of years more advanced than the human species, so why bother?

I submit that the real reason behind Drake's reluctance to have his model compete against Sagan's model in a fair, science-based, side-by-side comparison is because he knows that the Sagan Model would prevail. He is not willing to admit that Sagan was right and that he was wrong.

Yet, competition between hypotheses is fundamental to the scientific process. Carl Sagan required unsentimental appraisal of his hypotheses and those of any other scientist, demanding that all "prove [their] case in the face of determined, expert criticism."[7]

Frank Drake declares the Sagan Model dead on arrival by personal fiat rather than by open debate and rigorous investigation. He told Lee Billings: "Interstellar travel, on the other hand, I've worked on that quite a bit."[8] Drake uses the pronoun *I* rather than the collaborative *we*. He appears to have worked on interstellar spaceflight by himself, as an iconoclast, not as part of a scientific team, and then he declares the results of his research unequivocally true. If Drake is so cocksure that he's right about advanced aliens not being able to physically visit our planet, then why doesn't he allow the Sagan Model a chance to compete against his model in a fair and open venue? Why is he so isolated in his opinion? How come there isn't a bevy of space scientists standing with him, insisting that for all time and eternity there will never be anyone—humans or aliens—engaging in interstellar spaceflight? Where have all of his friends gone?

A Matter of Relevance

Perhaps one reason why thoughtful people have abandoned SETI, other than for its failure to detect a signal, is because the Drake Model, unlike the Sagan Model, is not central to the question of whether or not there is extraterrestrial intelligent life in the Universe. Yes, if SETI had intercepted an alien signal, it would have confirmed that aliens exist, but from SETI's failure, one cannot conclude that they do not exist. The SETI experiment occupies just a small niche in a broad spectrum of possibilities. As Carl Sagan, Stephen Hawking, and others have pointed out, there are a number of good reasons why an advanced alien civilization might choose not to transmit radio signals.

In fact, it may not be a good idea for humans to transmit radio signals. That debate is ongoing today concerning a proposed project known as "Active SETI" that, if implemented, would convert SETI's radio telescopes into transmitters. A number of leading scientists, including Stephen Hawking, are opposed to humans sending messages into space.[9] Even Frank Drake is on record as saying it would be a waste of time.[10] If some of our brightest scientists think it a bad idea for humans to transmit electronic signals into space, why would hyper-intelligent extraterrestrials think otherwise?

The alternative to making remote contact with aliens by radio signals is direct contact by interstellar spaceflight. Carl Sagan knew that if there is extraterrestrial intelligence in our galaxy, then, statistically speaking, there is virtually no chance that they have not been to Earth. This is the crux of the Fermi Paradox and the Sagan Model, and the insight that SETI seems to have finally grasped in its full significance. This is why the Sagan Model deserves to be a core part of the modern equation and why the Drake Model is passé. This is why it is imperative that the Stanford Paper be revisited. From a theoretical perspective, not having it on the table for active consideration calls into question NASA's credibility as an unbiased scientific institution.

Let me be clear: Sagan's model doesn't belong to UFO or ancient astronaut enthusiasts. It belongs to Sagan. For a variety of reasons, the Stanford Paper has been lost and forgotten. Few have known of its existence, and my goal is to change that. In this 20-year anniversary of Carl's death and in the years to follow, my hope is that the Sagan Model will be openly discussed and debated in university classrooms, at scientific conferences, and in peer-reviewed papers all around the world, as people remember Sagan's extraordinary accomplishments and service to humanity.

NASA's current search for microbial life on exoplanets is simply another consideration in the equation. If there is simple life elsewhere, then, given the age of the Universe, there must also be intelligent life elsewhere. And if there is intelligent life elsewhere, then Earth should already have been visited. If NASA is successful in finding simple life forms, it will be compelling evidence that the Sagan Model is correct.

Whereas the Sagan Model is a necessary component that can't be removed without the central equation falling apart, the Drake Model is a peripheral consideration and can be eliminated without affecting the calculus. The NASA search for microbial life and past alien visitations to Earth are inextricably bound together. You can look at it two ways: negatively or positively. Negatively, if there are no aliens, then there is no microbial life, and NASA is wasting time and taxpayer money looking for it. Likewise, if there is no microbial life, then there are no aliens. Either way you look at it, if there are no microbial life and no aliens, we can conclude that we are alone in the Universe. Positively, however, if there are aliens, then there must be microbial life in space, and if there is microbial life in space, there must be aliens. And, if there are alien civilizations, we can conclude that some of them are long lived and have been to Earth. This circular chain of logic runs in both directions and it can't be broken: A number of these long-lived extraterrestrials may have stayed around for thousands of years, leaving scientifically verifiable evidence of their presence encoded in ancient manuscripts, including, possibly, the Old Testament, one of the books that Sagan specifically identified as perhaps holding the key to a discovery of unimaginable importance.[11]

True friends of Carl Sagan will not allow his memory to fade along with the sinking fortunes of SETI and radio telescopes. He deserves better than that. The world needs to hear the truth that Sagan was a brilliant theoretical scientist who posited that aliens have been to Earth. He was the only scientist in the West who openly endorsed the possibility of ancient alien visitations and the only scientist to build a conceptual framework accompanied by a distinct search strategy. If you are one of Sagan's millions of admirers, I urge you to join with me lobbying NASA, SETI, and professional skeptics to engage the facts openly and exhaustively and end the cover-up.

In addition, my hope is that serious individuals from all walks of life will have intelligent and rational discussions on the subject of ancient alienism around kitchen tables, in living rooms, at restaurants, and around the water cooler at work. From the credentials of the man who crafted it, to the Drake equation, to Fermi's Paradox, to the inexplicable rise of Sumerian culture and civilization, and so much more, Sagan's model of ancient alienism contains all the necessary components and meets all the high standards that science theorists and professional skeptics look for in a serious proposal.

Based solely on scientific merit, the Sagan Model was, and still is, clearly superior to the Drake Model. In a fair and neutral environment, it would have been Sagan's search strategy that would have been given priority status, with the radio telescope search supported as a secondary experiment. Instead, the Drake Model was selected without critical scrutiny, whereas the Sagan Model was cast aside like so much garbage. This terrible injustice needs to be made right, and the only way for that to happen is for NASA, SETI, and professional skeptics to confess the cover-up and recognize the Sagan Model as scientifically legitimate.

The $100,000,000 Nail

On July 20, 2015, Russian billionaire entrepreneur Yuri Milner issued a formal announcement. Through what he calls Breakthrough Listen, he generously pledged 100 million dollars over 10 years to the SETI Institute to complete its search for conclusive evidence of extraterrestrial existence using state-of-the-art radio telescopes and the latest computer technology. Milner made his YouTube announcement in front of a distinguished panel that included astrophysicist Stephen Hawking, astrophysicist Lord Martin Rees, radio astronomer Frank Drake, Carl Sagan's third wife, Ann Druyan, and astronomer Geoff Marcy. During the event, the name of Carl Sagan was mentioned several times—but, of course, not a peep about his ancient alien theory.

Personally, I think $100 million is a lot of money to spend for what will be the final nail in the Drake Model coffin. Milner surely knows that Sagan was the son of Russian immigrant parents; that he shared information about extraterrestrial existence with Russian scientists at the height of the Cold War; and that several top Russian scientists, who were as much pioneers in SETI research as Sagan and Drake, were sympathetic, and some even enthusiastic, to the idea of past alien visitations. The Western parochialism that decreed that advanced aliens would be incapable of physically reaching Earth never reared its ugly head in the Soviet Union the way it did in the United States. The Russians knew better.

Putting a significant damper on the announcement of Milner's generous gift, Martin Rees stated that the odds of a successful intercept were low. Hawking followed that up by saying that a SETI failure would not prove that aliens do not exist. Drake then chimed in by hinting, quite strongly, that $100 million may not be enough money, and a decade may not be enough time, to get the job done. Capping off the negative refrain, Marcy stated that he would be willing to bet, not his house, but Yuri Milner's house, that the project would be successful. By this time, Mr. Milner had to be thinking to himself, "What in the hell have I gotten myself into?!"

But that wasn't all the bad news. Frank Drake informed the audience that recruiting top talent to SETI is almost impossible because (1) it's extremely boring work, and (2) it produces no scientific papers. This means that despite Milner's generous grant, the only way to attract gifted young scientists to the project would be to deflect much or most of the money away from looking for ETI to searching for the electronic signature of as-yet-undiscovered natural phenomena not related to ETI.

Within the space science community, it is now common knowledge that SETI's moment has come and gone. The experiment is over and it can't be revived, not even with the injection of a large amount of money. The excitement is currently in searching for signs of microbial life on exoplanets with space-based optical telescopes. Basically, what Milner was being told was that he was pouring his money down a rat hole—but that it would be happily accepted and spent anyway.

At the close of the news conference there was an opportunity for journalists to ask questions. Had I been there, I would have asked why advanced aliens, if they had a burning desire to use radio signals to inform the entire galaxy of their existence, didn't just flood all 10 billion frequencies of the radio spectrum with such powerful pulses that they would shout out in unison and with unequivocal certainty, "We are here!" If radio signals are their chosen means of communication, why have they made them so hard to detect? Even as he was about to launch Project Ozma on April 8, 1960, Frank Drake later reflected that his assumptions about alien technological development being on par with human development may have been underestimations, and that their systems may have in fact been significantly more powerful and able to project over vast interstellar expanses.[12] He could have, and he should have. It makes absolutely no sense to conduct a major experiment that is based on the flawed assumption that the radio telescope capability of advanced aliens is no better than mid-20th-century human technology.

Carl Sagan's position was: Why are we humans so naive as to think that we have to do all the heavy lifting?[13] If aliens are transmitting radio signals, we should be able to turn on a monitor attached to a radio telescope and see blips jumping off the chart. Why, like passing kidney stones, have we been straining so hard for so long—with no results? In the complete absence of any alien-generated radio signals, why isn't the proper assumption to conclude that hyper-advanced aliens don't communicate with radio signals? When will SETI come to its senses and openly admit that its radio telescope search strategy is a failure and should be abandoned?

The biggest contribution the SETI experiment has made to the advancement of science is that it has served to remind us of the truth of one of a basic dictum: that a single hypothesis is never as good as two or more hypotheses locked in a friendly, high-stakes competition.

At its commencement, the Drake Model was wrapped in all the hype and glitter of a Hollywood grand opening. Until the *Apollo 11* moon launch, nothing in the history of science could compare. It was an extravaganza that attracted the attention of the world. Everyone agreed that there was no better way to prove the existence of extraterrestrial intelligence than to listen for their radio signals with one of the marvels of modern science, radio telescopes. By Drake's own admission, his model put aliens and humanity on roughly the same level, and that made us feel good about ourselves. We were led to believe that aliens, if they exist, were not so far ahead of us technologically that there would be an unbridgeable gap between their species and ours. And, to whatever extent they might be ahead of us, they would build in the necessary accommodations to ensure that an emergent civilization like ours would be able to intercept and decrypt their message. No one except Carl Sagan was thinking about a technological and evolutionary gap so wide that the only way that their species could productively interact with our species would be for them to physically visit our planet and, through patient teaching over hundreds of years, incubate and cultivate the development of human civilization.

Never buying into conventional wisdom, Carl Sagan was sure that extraterrestrials delivered their message to Earth in person and did it in a way that was infinitely better than sending radio signals. If Yuri Milner should find it within himself to agree to fund an exhaustive Earth-based search, SETI's radio telescope experiment could finally be retired, and he will be credited with making a key contribution to the discovery of extraterrestrial intelligence. I have sent him a signed copy of this book along with a personal appeal for his help. Here's hoping he reads it.

Frank Drake was committed to refuting the possiblity of previous alien intervention and to dissuading us from the other potentiality of that idea: fear of a future violent invasion.[14] What Drake cavalierly dismissed as wrongheaded was a not-so-subtle slam at

Carl Sagan's science-based ancient alien theory. Today, the Sagan Model is hopefully on the verge of being accepted by mainstream science as a legitimate search strategy, and that it will replace an experiment that has ended in failure.

Denials

In 1964, at the first international gathering of scientists interested in in extraterrestrial research, held at Byurakan, Armenia, Russian astrophysicist Nikolai Kardashev proposed that there are three different levels of alien civilizations. What quickly became known as the Kardashev scale is almost as popular in ETI research circles as the Drake Equation. Those in Type I could harness the energy of their home planet. Those in Type II could harness the energy of their home star. And those in Type III could harness the combined energy of all the stars in their home galaxy. We have been told over and over that space travel at or near the speed of light would require energy levels too high to make it feasible. We were told that even if that velocity could be reached, slowing the spacecraft down would pose an additional set of insurmountable problems. As recently as in 2014, military specialist George Michael, in his book, *Preparing for Contact,* states that interstellar spaceflight would be impossible for any civilization on the Kardashev scale less than I; humans are currently classified as 0.[15] Michael suggests that it will be another 10,000 years before humans develop interstellar spaceflight capability.

It was a SETI lie about the impossibility of interstellar spaceflight that destroyed Carl Sagan's efforts to establish a science-based search for evidence of past alien visitations to Earth. Years later, when the Drake Model was yielding no results, it was another SETI lie that looking for an alien signal from space was like looking for a needle in a haystack that kept the radio telescope search going, when, by all rights, it should have been shut down and abandoned.

First, the anti-interstellar spaceflight lie, SETI lie #1: Interstellar spaceflight, whether for aliens or humans, is impossible. The Sagan Model of ancient alienism rests on one crucial assumption: that long-lived extraterrestrial civilizations would have had both the time and the technology to successfully traverse the vast distances of interstellar space to get from their home planet to ours. Yet, despite such mind-blowing accomplishments, NASA was still insisting that all aliens, even those in the Type III category, would be unable to reach Earth. Go figure.

I believe in 1964, when the Pentagon saw what Sagan was proposing, they conspired with NASA to bury his research. James Webb, the administrator at NASA when Sagan's paper came out, knew that he had the option of implementing both search strategies simultaneously. I'm convinced that it was an absolute deliberate act on Webb's part to ignore the superior science of the Sagan Model and select only the search strategy that had military utility.

At the same time, NASA had to be able to justify its decision to reject the Sagan Model by factually demonstrating why it was scientifically bankrupt. The problem was that the Stanford Paper was so extremely well thought out that it had no such glaring weakness—so NASA invented one. I believe that it launched an unofficial campaign that directed all NASA and SETI scientists to insist that interstellar spaceflight was physically impossible. No matter how much older and technologically advanced any alien civilization might be, they would argue that their ability to physically come to Earth was such a preposterous idea that it wasn't even worth discussing.

And now, the death blow, the coup de grace, is that NASA is busy with plans to build a working starship by the end of this century. Under NASA oversight and with government funding from the Defense Department, two starship-building initiatives, Icarus Interstellar and the 100 Year Starship Project, are currently working on meeting their expressed goal of constructing an operational interstellar spacecraft by the end of this century.

This is either one of the more amazing coincidences in the history of science, or a sure indication of a conspiracy and cover-up. Think of it: While Sagan was alive, talk of interstellar starships, even for advanced aliens, was anathema. Soon after his death, NASA starts building one. So why is NASA now working at building a starship by the end of this century? What in the hell is going on? Have we been lied to? The answer is obvious: yes, we have.

The 100 Year Starship Project makes the Apollo mission look like small potatoes, and yet it is advancing with minimal publicity. The general public has no awareness of what is going on. One reason for this institutional shyness is that it's clear that NASA wants the world to forget about Carl Sagan. There are still too many of his fans around who, if they knew, might begin asking NASA officials uncomfortable questions, such as why NASA rejected the Stanford Paper, in which Sagan wrote: "Especially allowing for a modicum of scientific and technological progress within the next few centuries, I believe that interstellar spaceflight at relativistic velocities to the farthest reaches of our Galaxy is a feasible objective for humanity."[16]

These words are prophetic, and under normal circumstances one might assume that NASA would print them on a banner and fly them high on a breeze. But NASA, in this new century and new millennium, would prefer that the name of Carl Sagan, along with Frank Drake, be relegated to the dustbin of history as co-founders of a pathetically unremarkable search program that ended in ignominious failure.

Unless it is trying to keep from drawing attention to its most famous scientist, one has to wonder why NASA doesn't call its interstellar endeavor the Carl Sagan Starship Project, after the man who pioneered the concept of human interstellar spaceflight. Or how about a Carl Sagan Starship Research Center? And why doesn't NASA reintroduce the Stanford Paper as proof to the world that Sagan, more than a half century ago, was suggesting that in just a few centuries NASA would be exploring the galaxy? Postmortem, NASA could use Sagan's writings and his fame to attract support for what will be by far its most ambitious undertaking ever. Only if there is a cover-up going on does it make sense for NASA to keep its starship project quiet and leave Sagan's name out of the mix.

Alien Advancement

Paradoxically, at the same time that SETI scientists and professional skeptics were telling us that the problems with interstellar space-flight were too great to be overcome, even by advanced aliens, those same scientists and skeptics were insisting that advanced aliens, if we were to ever meet them, would possess godlike powers and abilities far beyond our wildest imagination. For some strange reason, no one has ever bothered to collect and compare what the experts were saying about interstellar spaceflight against what they were saying about extraterrestrial genius and technological achievement. It is a study in contradictions.

For example, in the same book where Frank Drake writes that crediting extraterrestrials with scientific knowledge and technology much beyond our own would be "unfounded speculation," he states that he believes that advanced aliens would likely be immortal.[17] In other words, aliens are smart enough to have figured out the secret of extending life forever—but too dumb to have figured out how to physically reach Earth.

Another contradiction: Italian mathematician and SETI astronomer Claudio Maccone recently estimated that extraterrestrials a mere one million years ahead of humans would be 300 times more technologically advanced than modern humans are to mold spore![18] Yet, according to NASA, these hyper intelligent and hyper advanced aliens haven't been able to develop interstellar spaceflight capabilities.

I have to confess: For years I was listening to these contradictory narratives and not paying attention to the little voice in the back of my head that was saying, "This doesn't add up; it doesn't make any sense." I, like many others, was completely buffaloed by NASA scientists that I assumed knew what they were talking about—when they were making polar opposite statements.

Professional skeptics are equally guilty of perpetuating this conundrum. In what Michael Shermer calls his "Last Law," he compares advanced aliens to God. He writes: "For an ETI who is a million years more advanced than we are, engineering the creation of planets and stars may be entirely possible. And if universes are created out of collapsing black holes—which some cosmologists think is probable— it is not inconceivable that a sufficiently advanced ETI could create a universe by triggering the collapse of a star into a black hole."[19]

According to Shermer, advanced aliens might be able of creating new universes—but, amazingly, they haven't been to Earth!

Michael Shermer's friend and colleague, evolutionary biologist Richard Dawkins, apparently agrees with his Last Law. He expressed the God/alien Equivalency Principle with his typical unrestrained eloquence when he wrote: "Whether we ever get to know them or not, there are very probably alien civilizations that are superhuman, to the point of being god-like in ways that exceed anything a theologian could possibly imagine. Their technical achievements would seem as supernatural to us as ours would seem to a Dark Age peasant transported to the twenty-first century."[20]

Where did the God/alien Equivalency Principle originate? Shermer and Dawkins may have gotten it from Carl Sagan, who asked if the legends that describe alien beings with unimaginably powerful technologies would make them seem to the human population like gods.[21]

Incredibly, there are still SETI scientists and professional skeptics who, while extolling the godlike qualities of aliens, insist that there is no way that they could colonize the Galaxy. It's like two freight trains racing towards one another, full throttle, on the same track: unlimited alien capability on a collision course with an alien inability to reach Earth. The result is what one might expect: an ontological train wreck.

Sagan, in contrast, had absolutely no problem attributing intelligent extraterrestrial civilizations thousands or millions of years older than ours with knowledge and capabilities beyond our comprehension that would make it a near certainty that they have been to Earth. He maintained that if we were to physically meet such beings, they would seem godlike, which is precisely how primitive human civilizations came to describe their non-human visitors not long after their first encounters. The most common way of framing this argument is to ask where human science and technology might be in another hundred years, or another thousand years, or another million years. If we are prepared to credit our own species with a reasonable rate of progress over an extended period of time, how can we withhold it from aliens? Considering the overpowering logic of Fermi's argument, and the fact that NASA is currently at work building a starship, isn't it is a far safer assumption to think that advanced aliens have visited Earth than that they haven't? Isn't it then also a reasonable assumption to conclude that a search for alien artifacts, including those of a literary nature, would be a more logical search strategy than one based on the faulty argument that space is too vast for aliens to conquer?

Modern scientists tend to get puffed up with pride over what they have achieved, but the reason they get puffed up is because they compare their current scientific accomplishments with where science was a mere 400 years ago. What if, instead, they compared themselves to scientists 400 years in the future? If they did that, I think most of them would admit that they are still in diapers, barely learning to crawl. Now, what if they were to compare themselves to extraterrestrials who were one million years more advanced? If they did that, they might be forced to admit that modern science, as amazing as it is, is little more than a recently fertilized egg.

In his Stanford Paper, Sagan argues very persuasively that to appreciate what extraterrestrials might be capable of, all we have to do is measure where human technology is today compared to where it was a century ago, and how it has progressed from decade to decade. When we do this, we invariably find that technology advances at an exponentially ever faster pace, so that what was accomplished in the final decade of a century may amount to more than what was accomplished in the previous 90 years. This principle, known as Moore's law, is widely applied to advances in computer intelligence, but it is also persuasive evidence that Sagan was correct in arguing that alien interstellar spaceflight is a near certainty, and that the concept that aliens have visited Earth is scientifically plausible. Sagan wrote: "We desire to compute the number of extant galactic communities which have attained a technical capability substantially in advance of our own. At the present rate of technological progress, we might picture this capability as several hundred years or more beyond our own stage of development."[22]

When it came down to NASA having to decide how best to go about looking for evidence of extraterrestrial intelligence, the Drake Model or the Sagan Model, Sagan, guided by the established scientific principle of encouraging competition between contending hypotheses, was confident that intelligent and responsible leadership would have no choice but to implement both search strategies. The only way the Sagan Model would not be allowed to compete would be if there were a conspiracy to suppress Sagan work. In the end, as we now know, that is what happened. All the key players at NASA denigrated Sagan's research by insisting that alien interstellar spaceflight was impossible, leaving the poor bastards no choice but to communicate with us by radio signals. Now that the Pentagon and NASA are engaged in their own starship development, that lie has finally caught up to them.

Cognitive Dissonance

One of the favorite terms professional skeptics use in describing advocates of pseudoscience is *cognitive dissonance*, which Michael

Shermer defines as "the mental tension experienced when someone holds two conflicting thoughts simultaneously."[23] That Shermer, Dawkins, and other skeptics can actually believe that godlike aliens might be capable of creating new universes—and yet not have the ability to physically travel to Earth because the distance is too great—has to be the penultimate example of cognitive dissonance. Such an extreme and illogical stance is either the product of a very superficial thinker or the attempt on the part of a very cunning thinker to cover up Carl Sagan's work on past alien visitations to Earth.

Tracing this ruse back to its origins leads to the doorstep of the Pentagon. Not interested in any search program that lacked the potential for military application, it is obvious that insiders favored a search strategy involving radio telescopes that it knew wasn't able to compete on a level playing field against the Sagan Model. In 1964, when the Sagan Model was rejected, NASA knew full well that advanced extraterrestrials, if they exist, would not only own the technology to reach Earth, but that there was a high probability that Sagan was right in proposing that they were here.

The Sagan Model needs to be formally recognized and implemented. Now that the Big Lie #1 has been exposed, it's time for those of us who have been snookered by NASA into thinking that the distances between galactic civilizations is too great for advanced extraterrestrials to overcome, to express our righteous indignation.

SETI lie #2: Searching for an alien signal is like looking for a needle in a haystack. Carl Sagan advanced a two-pronged strategy for finding verifiable evidence of extraterrestrial existence. One was to employ radio telescopes to look out in space for an electromagnetic signal; the other was to look here on Earth in ancient manuscripts for a literary signal. SETI is an acronym that stands for the Search for ExtraTerrestrial Intelligence. In principle, it identifies a noble and worthwhile scientific goal without restricting itself to a single search strategy. Unfortunately, NASA chose to disregard this commonsense principle by looking only for a radio signal from space, leaving Sagan's Earth-based ancient alien strategy ignored and forgotten.

To ensure that there was no blind spot, Sagan endorsed both strategies—one that searched the heavens, and one that searched here on Earth. Though he clearly favored ancient manuscript research over radio telescopes, he thought it simple common sense to keep all options open. Still, Sagan was not shy about pointing out what he saw as many deep and serious flaws with radio wave messaging. His considered opinion was that the Drake Model was so suspect that he was extremely skeptical that trying to intercept an electronic signal would work. In the Stanford Paper, he writes:

> The difficulties of electromagnetic communication over such interstellar distances are serious. A simple query and response to the nearest technical civilization requires periods approaching 1000 years. An extended conversation—or direct communication with a particularly interesting community on the other side of the Galaxy—will occupy much greater time intervals, 100,000 to 1,000,000 years.
>
> Electromagnetic communication assumes that the choice of signal frequency will be obvious to all communities. But there has been considerable disagreement about interstellar transmission frequency assignment even on our own planet; among galactic communities, we can expect much more sizable differences of opinion about what is obvious and what is not. No matter how ingenious the method, there are certain limitations on the character of the communication effected with an alien civilization by electromagnetic signaling. With billions of years of independent biological and social evolution, the thought processes and habit patterns of any two communities must differ greatly; electromagnetic communication of programmed learning between

two such communities would seem to be a very very diffi-cult undertaking indeed. The learning is vicarious. Finally, electromagnetic communication does not permit two of the most exciting categories of interstellar contact—namely, contact between an advanced civilization and an intelligent but pre-technical society, and the exchange of artifacts and biological specimens among the various communities.[24]

Sagan then adds: "Interstellar space flight sweeps away these difficulties. It reopens the arena of action for civilizations where local exploration has been completed; it provides access beyond the planetary frontiers."[25]

Sagan's First and Last Radio Telescope Search

Remarkably, despite his misgivings about radio telescopes, Carl Sagan went on to participate with Frank Drake in the first large-scale galactic search for an alien signal. In 1975, "Sagan participated in the only SETI experiment of his career. . . . Sagan and Drake aimed the telescope at a large nearby galaxy. During the first half-hour alone, they scanned ten billion stars."[26]

But Frank Drake noted that Sagan seemed bored.[27] Why was Sagan bored? It was because he knew then that the radio telescope experiment was a failure. If no signal was detected after scrutinizing billions of stars, he was certain that humans would never capture an alien signal with radio telescopes, no matter how long they tried. If it didn't happen in that first hour, it wouldn't happen—not in the next 50 years, in the next 100 years, or in the next 1,000 years. Though by that time he was politically savvy enough not to announce his skepticism of the Drake Model publicly, Sagan abandoned radio telescopes as a viable search strategy. He knew in 1975 that they weren't the way to establish contact, and, as in so many other areas, he has been proven right.

Since that time, radio telescope SETI has continued on, growing ever larger and more costly, with tremendously greater search capacity—and no results. Despite a significant deterioration of influence and credibility, SETI Institute officials like Frank Drake are still reassuring people that electromagnetic contact with aliens is only a matter of time and that with more money and more sophisticated equipment it is sure to eventually succeed. The public has been reminded over and over that the SETI search for an electromagnetic alien signal is like looking for a needle in a haystack. This analogy is not only not true, but is beginning to wear thin among an aging but still adoring public that is desperate to know that aliens exist.

The excuse of proponents of the Drake Model that searching for an alien radio signal is like looking for a needle in a haystack is bogus, and they know it. For years it has been foisted on the public, and the public has bought into it. To the scientifically untrained ear, it sounds so reasonable. It encourages laypeople to think of the immensity of space and imagine SETI scientists courageously trying to intercept some faint, teeny, little radio signal emanating, almost unperceptively, from some distant alien planet. That is pure unadulterated bullshit. The reality is that if there is one alien radio signal in space, there have to be trillions of them. From the second the experiment first started, SETI radio telescopes should have been lighting up with successful intercepts.

SETI supporters who continue to call on the public for patience and financial support on the grounds that searching for an alien signal is like looking for a needle in a haystack, have no answer for critics like futurist Ray Kurzweil, who writes:

The SETI project is sometimes described as trying to find a needle (evidence of a technical civilization) in a haystack (all the natural signals in the universe). But actually, any technically sophisticated civilization would be generating trillions of trillions of needles (noticeably intelligent signals). Even if they switched away from electromagnetic transmissions as a primary form of communication, there would still be vast artifacts of electromagnetic phenomenon generated by all of the many computational and communication processes that such a civilization would need to engage in.[28]

Carl Sagan knew this, which is why he predicted from the beginning that radio telescopes were not the way to prove alien existence. The SETI Institute now hesitatingly acknowledges that Earth, by all rights, should have been visited by extraterrestrials eons ago, and it has all but given up hopes of intercepting an alien signal. The Drake Model is a scientific failure and SETI has become irrelevant. In the interests of scientific integrity, and without presuming that it will be any more successful than the Drake Model, the Sagan Model needs to be activated.

CHAPTER 4

A Conspiracy Wrapped in a Paradox

Quantify. What is vague and qualitative is open to many explanations.

—Carl Sagan, Baloney Detection Kit

For a full half century, SETI dodged and weaved around Fermi's Paradox, never taking it on head-to-head in a serious and comprehensive manner. But recently, in a quiet but historic change of heart, the SETI Institute revisited Fermi's Paradox, and it now openly admits that it is a virtual certainty that if advanced extraterrestrial civilizations exist, as the Drake Equation predicts, they would be capable of interstellar spaceflight, as the Fermi Paradox predicts, and, by all rights, some of them should have already been to Earth. This stunning change of doctrine unofficially brought 50 years of SETI's public denial of interstellar spaceflight to an end.

Following are selected excerpts from the SETI Institute Website essay on Fermi's Paradox. As you read, keep in mind that this is the same organization that for decades vehemently denied any possibility of interstellar spaceflight, either for humans or for extraterrestrials, a denial that turned Carl's Stanford Paper into a lost document:

Is there obvious proof that we could be alone in the Galaxy? Enrico Fermi thought so—and he was a pretty smart man. Might he have been right?

A lot of folks have given this thought. The first thing they note is that the Fermi Paradox is a remarkably strong argument. You can quibble about the speed of alien spacecraft, and whether they can move at 1 percent of the speed of light or 10 percent of the speed of light. It doesn't matter. You can argue about how long it would take for a new star colony to spawn colonies of its own. It still doesn't matter. Any halfway reasonable assumption about how fast colonization could take place still ends up with time scales that are profoundly shorter than the age of the Galaxy. It's like having a heated discussion about whether Spanish ships of the 16th century could heave along at two knots or twenty. Either way they could speedily colonize the Americas.

[Note: After introducing a several admittedly feeble explanations why we don't see aliens walking around on Earth that have been proposed by SETI theorists over the years, the essay continues.]

The presence of aliens on Earth would neatly solve the Fermi Paradox.

But while this (UFOs) is a prevalent idea among the public, the evidence for alien visitation has failed to sway most scientists. To convince researchers, who are inherently skeptical, unambiguous and repeated detection of flying objects by satellites or ground-based radar would be required. Better yet would be indisputable physical evidence, such as the landing lights from an alien craft. In other words, something better than witness testimony is necessary, since such testimony isn't good enough, no matter how credible the witness.[1]

Wow! What a turnaround! Step-by-step, the author walks us through Fermi's Paradox, acknowledging the full strength of its implications, just as Sagan did in his Stanford Paper. In fact, the reference to 16th-century Spanish sailing ships is straight out of the Stanford Paper, in which Sagan states: "The situation bears some similarity to the post-Renaissance seafaring communities of Europe and their colonies before the development of clipper and steam ships."[2]

When I saw this essay, I could hardly believe my eyes. For a half century, NASA and SETI had ignored the Sagan Model because it was based on alien interstellar spaceflight, a technological ability that it had deemed impossible. In one fell swoop, that impediment is erased by this essay, and Frank Drake's argument that space is too vast for aliens to colonize is completely and irreversibly shredded.

The logic of Fermi's argument is indisputable, but his conclusion that aliens don't exist was based on the complete absence of any reliable scientific evidence that they are here. Sagan thought that he knew where the smoking gun evidence that Fermi demanded might be located: in ancient Sumerian related manuscripts. Without in any way changing what Fermi was saying, Sagan simply expanded his argument to include the past. In the Stanford Paper he called for a sustained search of certain ancient documents, something that never happened because of NASA's across-the-board denial of interstellar spaceflight that flew in the face of Fermi's logic.

Physicist and science writer David Grinspoon, whose family had been close friends with the Sagan clan, writes about Fermi and his famous challenge to his peers in his delightfully informative book, *Lonely Planets*:

> However conservatively you work the numbers in the Drake Equation, it's hard to avoid the conclusion that we live in a widely inhabited galaxy, even if stars with living worlds are only one in a million. Fermi thought that, by this same logic, we should already have been visited. What if, in addition to developing radio technology for communications, advanced species also develop interstellar travel and decide to explore

or migrate to planets around other stars? Then, isn't a search for their presence in our own solar system just as valid as a radio search for their distant messages? How, then are we to interpret the fact that, as yet, we have found no scientifically accepted evidence for the past or present visitation of intelligent aliens? Can't we conclude that they do not exist and save ourselves the trouble of searching for signals?[3]

Note also how similar the SETI essay is to the description of Fermi's Paradox by David Grinspoon. This is not an accident. Grinspoon devotes an entire chapter of his book to Fermi's Paradox, and the SETI essay is basically a highly condensed version of what in my opinion is the finest and fairest analysis of the subject in existence.

The SETI essay on Fermi's Paradox is a historic concession that, by all rights, should have made front-page news in every newspaper and heralded by every media outlet in the world. For the SETI Institute to publicly acknowledge that it is a near certainty that if extraterrestrial civilizations exist, they would have the technical ability to engage in interstellar spaceflight and had more than enough time to reach Earth—is the most exciting and, in many ways, the most shocking development in its 50-year history. It is tantamount to a public concession that the Sagan Model of ancient alienism might very well be true. All that would be needed to conclusively prove that Sagan was right would be to discover the alien signal he predicted might be found in ancient manuscripts.

In this amazing essay, SETI does what Sagan did in 1962: link the Drake Equation with Fermi's Paradox with the 10-billion-year age of the Milky Way Galaxy. The Drake Equation is a compelling argument that extraterrestrials exist "out there." Fermi's Paradox is an equally compelling argument that states that if aliens are "out there," then they should be, or should have been, "down here." SETI, apparently, now agrees.

SETI's admission that alien star travel is highly likely runs against its original justification for conducting a radio telescope search that was premised on the false notion that interstellar space-flight is impossible for aliens and for humans. This was why the subject was never discussed at the Green Bank Conference, and why Sagan's Stanford Paper was rejected and abandoned without proper investigation and analysis. Now that interstellar spaceflight is considered a given, the question to ask is: If aliens could have and should have already visited Earth, why would they bother sending us radio signals? Fermi's Paradox makes the search for an alien signal with radio telescopes somewhat of an unnecessary afterthought. By accepting alien star travel as an existential reality, SETI has constructed a perfectly logical explanation for its failure to detect an electronic signal, and an equally compelling reason for it to open up an active investigation of the Sagan Model and proceed with haste to test any evidence that may proves that Sagan was right.

Go back to the SETI essay and begin with the sentence "The presence of aliens on Earth would neatly solve the Fermi Paradox."

This is a critical point in the essay. After effectively describing the full force of the Fermi Paradox, the author does something that I consider so disingenuous that it reminds me of tactics routinely employed in pseudoscience. He blunts the main thrust of Fermi's Paradox by identifying but a single alternative to radio telescopes: finding direct evidence of alien spaceships (UFOs). The author leaves no room and gives no opportunity for Fermi's Paradox to be solved in any other way except by these two means. He leaves the uninformed reader thinking that if direct evidence of UFOs isn't produced, the only fallback position is a continued radio telescope search for an electromagnetic signal. There's Strategy A and Strategy B, with no mention of Carl Sagan's Strategy C.

In this respect, the article is blatantly misleading, and departs from David Grinspoon's narrative, in which he writes: "How, then, are we to interpret the fact that, as yet, we have found no scientifically accepted evidence for the *past* or present visitation of intelligent aliens? [emphasis added]"

By specifically including the possibility of *past* aliens visitations as well as the possibility that they may be currently present, Grinspoon's interpretation of Fermi's Paradox is fair, accurate, and without bias. It leaves us to assume that Mr. Grinspoon would be in favor of testing concrete evidence that may confirm the Sagan Model because he has stated that he thinks it plausible that aliens may have visited Earth in past ages. Theoretical astrophysicists apparently now agree that if Earth has not been visited by aliens, our Galaxy must be the most pathetic and backward place in the entire Universe, and that the aliens who inhabit our Galaxy, if there are any, are little more than the cosmic equivalents of Harry and Lloyd in the comedy classic *Dumb and Dumber*. For the SETI essay to specifically *not* mention past alien visitations as a possible resolution to Fermi's Paradox is either an example of professional ineptitude or extreme prejudice. The author ignores Sagan's research and states that SETI would be interested in testing what someone claimed were UFO landing lights. In an obviously intentional and inexcusable act of neglect of formal scientific research, neither Sagan's name nor his Stanford Paper is mentioned.

We may never know why SETI chose to leave out the possibility of past alien visitations and focus solely on UFOs and radio telescopes as the only solutions to Fermi's Paradox. A cynic might suggest that it was a preemptive attempt to keep the Sagan Model out of the discussion. It may even have been because, when this essay was written, NASA and SETI were aware of the discovery of hard evidence that may confirm Sagan's ancient alien theory.

The author of the SETI essay, if he was fair and inclusive, should have added Sagan's search plan to the list as a Strategy C, but he didn't. The only solution to Fermi's Paradox that is offered in the essay as an alternative to SETI's radio telescope strategy is an outdated and totally debunked theory about UFOs that has been around longer than SETI has been in existence. Like SETI, UFO advocates have produced zero evidence, despite millions of ardent believers who would be quick to produce hard data if they had it. In effect, the author slipped in a red herring, a false positive that has been so thoroughly discredited that it has no hope of being revived. UFO theory, in effect, is SETI's official straw man. They keep propping it up and then knocking it down, each time making a radio telescope search appear, by comparison, all the more scientific and reasonable. This begs the question: How would the Drake Model stack up against the Sagan Model in a fair and impartial side-by-side competition?

By failing to mention the Sagan Model, even while it admits to the probability of alien interstellar spaceflight, SETI seriously distorts the facts while generating a smokescreen that keeps people from considering a scientifically credible option to both UFOs and SETI. What is most egregious is that in failing to mention the possibility of finding evidence of past alien visitation to Earth as a way of solving Fermi's Paradox, SETI, for the second time in its history, has slammed the door in Sagan's face, and, frankly, I find this lack of respect for a truly great scientist and even greater human being very disturbing.

Without fanfare, SETI added the Fermi Paradox essay to its Website, perhaps hoping that no one would notice the seismic shift it has created in SETI doctrine, a fundamental change that opens up stunning new possibilities that it had previously considered off limits. SETI's admission that alien interstellar spaceflight is a near certainty creates the perfect opportunity for it to publicly acknowledge its error in dismissing the Sagan Model as bad science. Instead, it has added insult to injury by not mentioning the existence of the Stanford Paper, or the possibility that if Sagan's theory of ancient alienism were proven true, it would solve the Fermi Paradox just as effectively as UFO landing lights. All I can say is shame on NASA, shame on SETI, and shame of any professional skeptics who may be complicit in this cover-up.

SETI's admission that interstellar spaceflight is possible is certainly commendable, but it can't take away the humiliation and suffering that Sagan experienced throughout his career for believing in alien star travel while SETI was doing its best to deny it. Neither can it buy back the 50 years that the Sagan Model has been secretly covered up by NASA.

At this point, there is no indication that NASA and SETI have any intentions of openly recognizing the Sagan Model of ancient alienism or in testing any evidence that supports it. This is not only a shame and an embarrassment for NASA and SETI, it is a terrible blow to the legacy of the man they openly honor as a pioneering genius in ETI theory.

Neil deGrasse Tyson

In spring 2015, the National Geographic television channel began advertising that it would begin airing Neil deGrasse Tyson's syndicated program, *StarTalk*. On Monday, April 20, with pen and paper in hand in case the subject of extraterrestrial existence came up, I made it a point to watch. Tyson's guest that evening was astrophysicist Charlie Liu, and the theme of the program was about the legitimate science behind the famous *Star Trek* series, and the adventures of the crew of the Starship Enterprise as they explored the Cosmos. It was the perfect scenario for the subject of aliens to come up—and, much to my delight, it did.

As Charlie and Neil bantered back and forth in conversation, they began to discuss the subject of whether humans will ever engage in interstellar spaceflight, and, if so, how many more lifetimes it might be before we get there. Neil is then asked if that day will ever come, and Tyson's response was "I don't see why not."

And then they got into what was clearly an awkward discussion for Tyson. Charlie began talking about eventual human colonization of the galaxy as our species "goes where no man has gone before," and spreads our DNA throughout the Cosmos like they did on *Star Trek*. Neil tries to change the subject by stating that anything like that could not happen in a single lifetime. But Charlie is persistent, and Neil agrees that it could be done, not in a human lifetime, but as a species. Charlie then adds that there is still plenty of time for us, as a species, to spread out and go to the stars, and Tyson builds on that argument by noting how bacterium in a petri dish will inexorably divide, multiply, and spread—until the entire solution is "colonized."

At that moment, Tyson says, "Something like Fermi's Paradox." After the break, they moved on to another subject.

What may have been Tyson's Freudian slip of the tongue in mentioning Fermi's Paradox in the context of human space travel is that if it is probable that human civilization will one day travel to the stars, highly evolved aliens who began exploring the galaxy millions of years ago should have already been to Earth. It apparently dawned on Tyson that Charlie Liu had walked him into a catch-22 situation where the audience might quickly deduce that, from both a logical and a scientific perspective, the odds that aliens, if they exist, have not been to Earth, are virtually nil.

The subject on *StarTalk* the following week was also about the possibility of human interstellar spaceflight, as Tyson and cosmologist Janna Levis discussed the science behind the movie *Interstellar*. Clearly, interstellar spaceflight is a subject that strikes a common chord among the public. With such broad worldwide interest, I suspect that Frank Drake and other NASA scientists who still vehemently deny advanced aliens that capability will not be invited as guests on the show anytime soon.

Many NASA scientists, including current administrator Charles Bolden, envision the possibility that humans might someday be able to build a Starship Enterprise. In fact Tyson even said how much it would cost: $46 billion. But NASA is not willing to attribute a Starship Enterprise to aliens because that would make the Drake

Model quantitatively inferior to the Sagan Model. Why would aliens send radio signals to Earth when they can get here physically?

For Carl Sagan, there was nothing at all paradoxical in Fermi's observation. Sagan calculated from math, from logic, and from the historical record, that aliens have been to Earth. The science is on the side of Enrico Fermi, Carl Sagan and ancient alien theory, not on the side of NASA, SETI, and professional skeptics. Those who continue to argue against the latest science by using the completely discredited argument that the distances between inhabited worlds is too great for any species to traverse, no matter how long lived and technologically advanced they might be, are either delusional or disingenuous.

A Preponderance of Evidence

In the early 1960s, NASA was presented with two ETI search strategies, one space-based and the other Earth-based. It had a choice. It could choose one or the other, or it could implement both. Based purely on the science, the Sagan Model was demonstrably superior, making the odds of it succeeding much higher than the odds of intercepting an alien signal with radio telescopes. Yet NASA went with Frank Drake and radio telescopes, which left them with the problem of what to do with the Stanford Paper. With complete arrogance and disdain for the scientific method, NASA administrator James Webb selected the Drake Model and then, I believe, went to great lengths to suppress and conceal the Stanford Paper.

At the time this momentous decision was made, NASA was a fledgling government bureaucracy led by individuals with high aspirations to see it grow into a classic government institution with a multi-billion-dollar budget, tens of thousands of employees, and huge political power and influence. NASA's first administrator, Dr. T. Keith Glennan, was a scientist who had been president of Case

Institute of Technology. There is little doubt that he would have appreciated the value of competition in the search for extraterrestrial intelligence. The second NASA administrator, James Webb, was in charge at the time the Stanford Paper and ancient alien theory were up for consideration. He is described on NASA's Website: "As a longtime Washington insider, he was a master at bureaucratic politics."[4] This is compelling evidence that Webb, not a scientist, was, in fact, a Pentagon plant who was under orders to vet every conceivable project that had anything to do with space, including the search for extraterrestrial intelligence, to make sure that it had the potential to be militarily useful.

Webb had no qualms violating the fundamental scientific principle of pursuing excellence by encouraging competition. He was confronted with two search strategies. One had military value, the other didn't. One created jobs for NASA astronomers, the other didn't. One would enlarge NASA's budget, the other wouldn't. Before either plan was officially accepted, Webb picked a winner and a loser. Drake won, Sagan lost.

Carl Sagan's Stanford Paper, with its claim that any truly advanced aliens would have already been to Earth, was considered a disaster at NASA. The Sagan Model would require NASA to hire archeologists, anthropologists, historians, and other academics who had no connection to space, which would have forced it outside of its institutional mandate. This was not only a distasteful prospect for NASA, but, if implemented, it might have raised questions in Congress about a NASA overreach.

Yet, all the Sagan Model amounted to was a nuanced reiteration of Enrico Fermi's comment that if advanced extraterrestrials exist anywhere in our galaxy, they should already have been to Earth. Fermi may have created the tiger, but it was Carl Sagan who gave it teeth. Sagan is now gone, but his tiger is back and it's biting NASA in the ass. After more than a half century, the Sagan tiger is finally out of its cage and on the prowl.

Carl Sagan's Stanford Paper put NASA at a crossroads. It could either stay true to its commitment to science and the scientific method, and recommend a two-pronged search strategy, *or* sell its soul to the military-industrial complex and go exclusively with a less-rigorous search strategy that was compatible with its mandate and stated sphere of interest. In crafting the Stanford Paper with the help of a NASA grant, the young and brilliant Carl Sagan had created a NASA nightmare, and it put him so deep into James Webb's doghouse that he would never get out.

NASA chose the Morrison/Cocconi Paper over Sagan's Stanford Paper. It was a deliberate and calibrated act of pure political expediency that defied and violated the scientific ethic and left Carl Sagan's 10-year investment in ancient alien research out in the cold. NASA leadership chose the single search strategy that they knew would bring it huge sums of government money and keep its astronomers gainfully employed for decades—over a scientifically superior strategy that would have shifted funding and notoriety away from NASA and direct it to academic disciplines that had nothing to do with space exploration.

Carl Sagan, as one of America's brightest young space scientists, was one of the astronomers responsible for the founding of NASA in 1959. One can only imagine the shock in the embryonic NASA establishment when, three years later, Sagan introduced a theory in which he claimed that an Earth-based search for evidence of extraterrestrial existence would be far more likely to succeed than a space-based search.

Normally, when one scientist attacks the work of another scientist that has been peer reviewed and published, it is imperative that it be a critical and respectful analysis that addresses specific issues in microscopic detail. But the outrage against Sagan and his Stanford Paper was so palpable at NASA that its scientists were allowed—even encouraged—to attack him with impunity with such off-the-cuff epithets as "bad science"[5] and "something that you might read on the back of a cereal box"[6] without fear of reprisal. It was open season, and Sagan was in everyone's crosshairs.

James Webb's fateful decision to go with radio telescopes and, perhaps, to cover up Sagan's ancient alien research was so successful that today, when people think of the search for extraterrestrial intelligence, they automatically think of radio telescopes. Conversely, those same people, when they think of ancient aliens, automatically think of junk science and scam artists. From a historical perspective, NASA owns the ETI genre lock, stock, and barrel. Carl Sagan, the Stanford Paper, and his ancient alien theory are nowhere to be found on NASA's radar screen, and every Sagan biographer I have read has either failed to mention the Stanford Paper or buried it in obscure verbiage.

Cause for Concern

Committing the search for extraterrestrial intelligence to an exclusively space-based enterprise, although it may have served Pentagon interests and help build NASA's organizational pyramid, was putting it on a trajectory that was doomed to failure. Now that the radio telescope experiment is all but dead, one would think that NASA would have no choice but to reconsider ancient alienism as a viable alternative, but that isn't the case. Like the weird family that keeps Grandpa's lifeless body in a living room rocking chair so they can continue to cash his social security checks, NASA is hesitant to publicly admit the obvious, that the radio telescope experiment has failed and the corpse is beginning to stink. Even worse, it appears that NASA is prepared to keep SETI's lifeless body artificially propped up in perpetuity, evidently thinking that as long as it is not formally abandoned, the chances that someone will rediscover Sagan's work on ancient alienism will be minimal.

The era we live in represents but a small slice of the approximately 10,000 years that humans have been civilized. What would prompt SETI to think that claims of present-day alien activity are worth considering, but not claims of past alien visitations? Why would SETI express a willingness to test physical evidence of UFOs, but not physical evidence left behind by aliens who may have been on Earth thousands of years ago for extended periods of time and have since left?

Considering that NASA is currently working on interstellar spaceflight, it is entirely reasonable to think that advanced aliens might have visited Earth in the past and then moved on. Why is it necessary to prove that Extraterrestrials are here on Earth today to prove they exist?

The SETI Institute needs to clarify itself on this issue. Assuming the underlying theory is sound and the evidence is empirical, why would it not test data that was discovered in an ancient manuscript? Why is it only landing lights that dropped off an alien spacecraft that SETI is willing to examine?

WWCD (What Would Carl Do)?

On the June 15, 2015, episode of *StarTalk*, Neil deGrasse Tyson's featured guest was NASA administrator General Charles Bolden. He predicted that humans would be on Mars in about 20 years, and then he went on to say that he can envision a time, far off into the future, when humans will follow the *Voyager* spacecraft and travel beyond our solar system—exactly what Carl Sagan proposed in the Stanford Paper. And, if humans will someday do that, why couldn't advanced aliens, thousands or millions of years ago, have developed the technology to reach Earth? Sort of makes sense, doesn't it?

If proving the existence of extraterrestrial intelligence is as important as everyone thinks it is, then why leave any credible search strategy unattended or any credible evidence uninvestigated? Carl Sagan's model of ancient alienism and new discovery evidence that supports that model are on the shelf, waiting to be properly vetted by qualified experts. On this 20-year anniversary of Carl Sagan's death, if NASA, SETI, and professional skeptics continue to deny that Carl Sagan was an ancient alien theorist, who is going to step forward and honor Sagan's legacy in a truthful manner? Setting the record straight may require a mass petition by Sagan's millions of adoring fans. If we insist that the Stanford Paper be extended the attention it deserves, and that all serious research on the subject of ancient alienism be openly recognized, perhaps we can foil Pentagon hopes that the world will forget about Carl Sagan.

Establishing Attribution

Some might say that if the Sagan Model is ever confirmed, the credit should go to Enrico Fermi, not Carl Sagan. After all, as early as 1943, Enrico Fermi was floating the idea among his peers that if extraterrestrials exist anywhere in our galaxy, they should have already been to Earth. So wouldn't he deserve credit for the theory based on having made the prior claim?

The answer is no, because Enrico Fermi was fully occupied in important research unrelated to extraterrestrials, and he limited his involvement to that of an anecdotist. The fact is, Enrico didn't have time to be an ETI theorist. Carl Sagan picked up on Fermi's casual anecdote and put flesh on the bones. Sagan became the world's first ancient alien theorist in the full scientific meaning of the term. Fermi may have primed the pump, but it was Sagan who delivered the water.

Like everyone else who has ever entered the woods hoping to bag a deer, an elk, or a bear, when hunting season came along I would constantly be looking to the ground for fresh tracks that would indicate that the game I was stalking was nearby. And, as every hunter has experienced, sometimes there would be no fresh tracks, only lots of older ones, evidence that the quarry, though it had once been in the area, had since moved on.

When Fermi made his offhand remark he was evidently think-ing only about aliens leaving fresh tracks, not aliens from the past who may have left tracks that they had once been to Earth but had since moved on. For Carl Sagan, the ultimate alien hunter, look-ing for signs of past alien visitations to Earth was a natural thing to do. He knew that there were no aliens on Earth today, but when he looked into the historical record he saw unmistakable signs that they were once here.

Sagan's expanded interpretation of Fermi's thesis to include the past as well as the present created a second Earth-based search strat-egy that, in theory, complemented the UFO movement in much the same way that it complemented the Drake Model. All three had the same goal—to prove the existence of extraterrestrial intelligence—but each had a different way of reaching that goal. While Frank Drake was looking for alien radio signals and UFOlogists were looking to the skies for flying saucers, Sagan thought it reasonable to investigate ancient manuscripts for signs of aliens past, and he found them.

Sagan knew that it would require a professional and thorough investigation of manuscripts related to the Sumerians before his model of ancient alienism could be dismissed. He resisted the efforts of some of his peers who tried to discredit his research by using the "guilt by association" ploy that attempted to lump his model in with UFO theory. It doesn't work; the Sagan Model is a completely differ-ent kind of animal. Modern professional skeptics might try the same tactic by lumping his model in with tabloid ancient alienism, even though the differences between the two are as stark as night and day.

Having dismissed UFOs, but still acknowledging the strength of Fermi's observation, Sagan refined his statement by calculating that the odds that aliens have been on Earth in the past and have since moved on, are the same as the odds that they exist. In other words, if long-lived hyper-intelligent and hyper-advanced extrater-restrials are real, as the Drake Equation predicts, then it is a near scientific certainty that they have been to Earth—and, conversely, if, after an exhaustive search, no evidence is found that they have been to Earth, it is reasonable to conclude that they do not exist, so why bother with a radio telescope search, or, for that matter, with any other kind of search?

Carl Sagan built a solid theoretical base for ancient alienism so that credible scientific institutions could be ready to test and analyze any concrete evidence that might be discovered that had the potential to confirm his theory. We now have the theory and we know where to look for the evidence. All that is needed is for a credible scientific organization to test the data and release the results.

Impossible or Inevitable?

In his recollection of the 1961 Green Bank Conference, Frank Drake was of the opinion that the reason the subject of alien visitations to Earth was not discussed was because everyone there thought that interstellar spaceflight, either for aliens or for humans, was impossible. That extreme anachronistic position is now thoroughly falsified. In one of the latest books on the subject, *Beyond* (2015), University of Arizona professor Chris Impey concludes that the only thing that can stop humans from developing the technology to explore interstellar space within the next couple of centuries is self-annihilation. He writes:

> The 100 Year Starship project is funded by NASA and the Defense Advanced Research Projects Agency (DARPA). In 2012, a million dollar grant was awarded to former astronaut Mae Jemison and the non-profit organization Icarus Interstellar, for work toward interstellar travel in the next hundred years. It's important to realize that the majority of the speculative research on interstellar travel is being undertaken by professional physical scientists and engineers, with the work published in scholarly journals and books.[7]

I have to ask: How is it humanly possible for a modern scholar to write a book about interstellar spaceflight and not mention Carl Sagan's Stanford Paper? Incredibly, Impey even discusses the Bussard ramjet as a possible means of propulsion—the same system that Carl Sagan mentioned in his Stanford Paper in some detail. Writing at length about interstellar spaceflight, either pro or con, and not referencing Sagan's 1962 groundbreaking research on the subject is perplexing. One possible explanation is that it would be almost impossible to credit Carl Sagan with being one of the world's first interstellar spaceflight theorists—without also revealing that he was an ancient alien theorist, a historical fact that NASA doesn't want divulged.

Interestingly, Impey discusses Fermi's Paradox, but appears to buy into the NASA line that the only solution to it is if someone bags a UFO. He makes no mention of the fact that it can be solved equally well by finding conclusive evidence in an ancient manuscript that aliens have been to Earth in past ages and have since left.

At the same time, it's hard to blame any serious science writer for not wanting to broach the subject of ancient alienism. The waters of that subject have been so polluted by tabloid blarney that for a credible science writer to even mention the words "ancient alien" in his text as a reasonable thesis is to run the risk of being tarred and feathered by the mainstream scientific establishment. The question must be asked: Who is to blame for this sad state of affairs? Why hasn't the study of ancient alienism been recognized as an academically legitimate and science-based discipline? The blame lies with NASA, and it is NASA that has the power to change the situation by taking the lead and encouraging and supporting rigorous and disciplined research on the subject.

A NASA Conspiracy?

In my estimation, NASA's suppression of the Stanford Paper was a conspiracy that destroyed the legacy of its most famous scientist, so let's break it down. To have a conspiracy, you need five things:

knowledge, motive, means, execution, and cover-up. Let's examine each of these in regard to both allegations.

Knowledge

NASA provided the original funding for the Stanford Paper and it is preserved in its official archives, so its knowledge of Sagan's ancient alien theory is undeniable.

Motive

The Sagan Paper was an alternative ETI search strategy that didn't have military value, and, if implemented, would not have brought government money to NASA and would not have created jobs for astronomers. NASA, under orders from the Pentagon, buried Sagan's research before it saw the light of day.

Means

NASA had absolute control over Carl Sagan's career as a space scientist. They had the power to boot him out and they had the power to turn him into a superstar. Trapped in a coercive environment, Sagan had little choice but to capitulate and to give up on his dream of seeing the ETI search platform he developed become a reality.

Execution

After the Stanford Paper was written, it was peer reviewed and published in a respected scientific journal. That was the moment when it should have been critically scrutinized and critiqued by other scientists and academics in formal papers, some of which would have likely been published in scientific journals. Instead, it was as if the Stanford Paper had never been written. There are no recorded interactions of any kind, and, even though it was a NASA product, there is no record of any NASA response or analysis. For the formal research of a respected scientist who was openly acknowledged to be a leading expert in his field to be so universally ignored and not allowed due process is unprecedented in the annals of science.

Cover-Up

From the time the Stanford Paper was written, any mention to NASA about past alien visitations to Earth was met with the instant response that interstellar spaceflight, either for humans or extraterrestrials, was physically impossible. In a classic example of group-think, interstellar spaceflight was considered such a laughably implausible idea that it was considered not worth talking about. This broad-based denial effectively bamboozled the public into believing what we now know is a myth. Can any NASA scientist keep a straight face and insist that long-lived aliens would not have developed the technology to reach Earth—when NASA is actively engaged in developing an interstellar starship by the end of this century?

~~~

Does this add up to proof of a 30-year conspiracy? I think so, but I'll leave it for readers to decide for themselves. At the very least, an unholy and misguided triumvirate of NASA, SETI, and professional skeptics have some serious explaining to do. Why has the Stanford Paper been suppressed, and why hasn't the truth about Carl Sagan's belief in ancient aliens been told? The best that those of us who care about Sagan's legacy can do is to seek to repair his reputation and fight for the restoration and implementation of the ETI model he developed.

There are a lot of very blunt questions that need to be asked, and NASA, SETI, and professional skeptics need to respond fully and honestly. If they don't, they risk being implicated for attacking Carl Sagan's character and competence. Now is the time for NASA, SETI, and professional skeptics to speak out and defend themselves. In the light of the discovery of new facts in the historical record, for them to remain silent would be self-incriminating. Today, almost 60 years after it began, the NASA conspiracy to keep the public from learning that Carl Sagan believed in ancient aliens is still in force.

CHAPTER 5

SETI and Ancient Alien Wars

The statistics presented earlier in this paper suggest that the Earth has been visited by various galactic civilizations many times (possibly 10,000, more or less) during geological times.

—Carl Sagan

The disclosure that Carl Sagan was a lifelong believer in ancient aliens and crafted a science-based model is certain to have a major impact on the ancient alien community, and it is likely to be highly disruptive. No ancient alien theorist, living or dead, has scientific credentials that can compare with Carl Sagan.

Sandwiched between popular ancient alien founder Erich von Däniken on the one hand, and professional skeptic and anti–ancient alienist Michael Shermer on the other, is Carl Sagan, who was both an ancient alien theorist and a founder of the modern skeptic movement. As a skeptic, Sagan fought tirelessly against the von Däniken Model because he didn't believe that the evidence von Däniken cited measured up to scientific standards. At the same time, as an ancient alien theorist, Sagan argued in research that has been suppressed by NASA that it was aliens who introduced civilization to humans through the Sumerians.

Sagan himself said that every legitimate science has its pseudo-scientific counterpart. What this implies is that as the Sagan Model of ancient alienism gains traction, it will occupy the empirical end of the scientific spectrum, leaving the von Däniken Model to remain where it is: at the opposite end of the spectrum where it is broadly dismissed by NASA as pseudoscientific. For ancient alien enthusiasts, there is now a choice to be made: to stay with conventional advocates like Erich von Däniken, or to follow Carl Sagan.

This could create a schism within the ancient alien movement that, if it were to happen, would result in two opposing ancient alien camps, one where traditional scholarship is recognized, and claims are advanced and defended only to the degree that they can be supported by credible data; and the other camp, where traditional scholarship is routinely ignored and the bar for what constitutes supporting evidence is set extremely low.

In this scenario, those who identify with the Sagan Model will seek to interface with mainstream science and the mainstream media, while alternative ancient alienism will continue to spread its message through tabloids, esoteric books, slick Websites, late-night radio, and highly scripted television documentaries.

Let me take a moment to address both von Däniken and Shermer, and their millions of fans and followers. Von Däniken first.

The distinctions between the Sagan Model of ancient alienism and the von Däniken Model are philosophical, not granular. The Sagan Model is theory-driven and the von Däniken Model is evidence-driven, so it is simply not possible to draw any detailed comparison between the two. Sagan's strategy was to establish the science of interstellar spaceflight and then argue, based on Fermi's Paradox and ancient myths, that there is a strong likelihood that long-lived advanced extraterrestrials have visited Earth. That is where he stopped. His goal in establishing an Earth-based SETI to complement the radio telescope search was so that more research could be invested into the science of interstellar spaceflight and more investigations of ancient manuscripts that referenced encounters between humans and godlike beings who came from another place in the cosmos.

In contrast, the von Däniken Model of ancient alienism is based on specific archeological and historical evidence gathered from all over the world. Though it can appear impressive and convincing to non-scientists, establishment scientists and scholars haven't bought into it. They accuse von Däniken and other conventional ancient alien theorists of not following the scientific method. Except for an occasional joke and disparaging remark, scholars and scientists seldom mention ancient alienism, leaving the subject to professional skeptics such as Michael Shermer, who attack ancient alienism and its proponents with unrelenting vigor. Still, on at least one occasion, Sagan offered his opinion—at the centennial of the famous Gifford Lectures in Scotland in 1985 that was published as a book in 2006. Following are relevant excerpts:

> The fundamental hypothesis of von Däniken was that there is impressed in the archeology and folklore and myth of many civilizations on Earth certain indications of past contact with the Earth by extraterrestrials beings. This is not an absurd proposition on the face of it, but how acceptable the hypothesis is depends on how good the evidence is. And, unfortunately, the standards of evidence were extremely poor, in many cases, non-existent.[1]

Sagan then goes on to debunk, with surgical efficiency, von Däniken's theory that aliens helped the Egyptians build the pyramids, and that they assisted in designing the large animal drawings on the Nazca plains in Peru; and in creating the stone monoliths on Easter Island.[2]

In response to those he thought were making extraordinary claims about the Egyptian pyramids based on shaky evidence, Sagan frequently blended science, history, and logic with humor and a tinge of sarcasm. After citing the Greek historian Herodotus's writings about Egyptians transporting blocks of stone on rafts and moving them along land on rollers, he adds:

> There were even inscriptions on a few key blocks that say the equivalent of "My goodness, we did it!" signed "Tiger Team Eleven," which seems an unlikely delight in modest construction by some being who had effortlessly travelled through interstellar space. And we know that the first pyramid that was ever constructed fell down and that the second pyramid, halfway through construction, had the angle of the sides dramatically pared, because they had learned from the example of the first one that fell down.[3]

We have all marveled at the extremely accurate drawings of a hummingbird and other animals etched in the vast drawings on the Peruvian plains, and wondered how primitive humans could have pulled off such a feat without extraterrestrial assistance. Sagan, with no small amount of satire, compares this to fans holding up signs and doing the wave at sporting events.[4]

How could primitive people on Easter Island move blocks of monolithic stone images weighing several tons over long distances and then set them upright so they were looking out to sea? Conventional ancient alien theorists insist that they got help from alien visitors, but Sagan had a better answer: "Years before von Däniken wrote, Thor Heyerdahl with a small team . . . had transported and erected one of these monoliths that had been found in a supine position. And the erection method included just shoving small bits of dirt and stone under one side of it until it got to the high, steeper angle and then finally stood up."[5]

Scientists introduce claims and supporting evidence to their peers, inviting their critical analysis. These claims are often rebutted by counter-explanations of the evidence. Like an extended tennis match, these collegial exchanges and interactions can go on for years before a consensus is finally reached. Unfortunately, this traditional back-and-forth rarely happens with conventional ancient alien proponents, and it didn't happen in this instance. Erich von Däniken proposed, Carl Sagan disposed, and then von Däniken, rather than openly addressing Sagan's alternate explanations, disengaged from the conversation. The only conclusion one can draw is that von Däniken, without a logical comeback, chose to bypass mainstream science, ignore Sagan's rebuttal, and plead his case to the masses, who he knew would be less critical.

He then continues:

> I've presented some of his best cases. Fundamentally, what von Däniken has done is to sell our ancestors short. . . .
>
> I consider it an extremely dangerous doctrine, because the more likely we are to assume that the solution [to Earth's problems] comes from the outside, the less likely we are to solve our problems ourselves.[6]

For current followers of ancient alienism—and they number into the tens of millions worldwide—the discovery that Carl Sagan was a die-hard believer in ancient aliens and the recovery of his seminal research on the subject is a decisive moment. Will they continue to follow the ancient alienism leaders they have in the past, or will they switch over to the Sagan Model and help fulfill Sagan's dream of seeing ancient alienism become a part of the scientific mainstream, which, in my opinion, is where it deserves to be?

For 30 years I have been a devout believer in ancient aliens—but, by intention, I chose not to identify with Erich von Däniken and the mainstream ancient alien movement. Instead, I identify myself with Dr. Carl Sagan, who, more than 10 years before Erich von Däniken wrote *Chariots of the Gods?*, was openly telling close friends and associates that he was convinced that Moses and Jesus were extraterrestrials. The following excerpt is from the Sagan biography *A Life in the Cosmos*, by William Poundstone:

One Sunday morning toward the end of summer, Abrahamson, his fiancée, and Sagan were washing the car the couple had been given as an engagement present. Sagan propounded a new theory: that Moses, Jesus, and all the great religious figures of ages past were really extraterrestrial beings. The miracles of the Bible had all happened as described. Moses parted the Red Sea, Jesus turned water into wine, and so forth. They used advanced technology that was perfectly normal on their planet—but which we earthlings could take only as proof of divinity.

Abrahamson good-naturedly challenged him. Sagan refused to budge. He was either serious or acting as if he was serious. With Sagan, it was hard to tell which.

That afternoon, Abrahamson took his fiancée and Sagan out to dinner at what was, for Bloomington in the 1950s, a very posh establishment. It was "the kind of restaurant where people went after church." In the middle of dinner, without any warning, Sagan slammed his fist on the table, sending the dishes rattling. He looked Abrahamson in the eye and bellowed, "I tell you, Jesus Christ is extraterrestrial!"

The restaurant fell silent. It took a subjective eternity for conversations to resume with something of their former spontaneity. Abrahamson and his fiancée wanted to crawl under the table.[7]

With all due respect to Erich von Däniken, Zecharia Sitchin, and other ancient alien theorists of the modern era, the question needs to be asked: Why tether one's belief in ancient aliens to them, the Johnny-come-latelies, when one can follow the ancient alien beliefs and teachings of one of the most accomplished astronomers and celebrated scientists of the 20th century?

With the discovery of Carl Sagan's science-based model, ancient alien aficionados now have the opportunity to step out of the shadows and become part of the scientific mainstream. It's an opportunity for them to associate themselves and their cause with one of the most respected scientists of the modern era. Why would they not welcome that opportunity?

One possible reason is that, if they choose this path, it will mean disassociating themselves from popular ancient alien leaders and organizations that they have come to know and trust. Another reason is that they may have a hard time adjusting to Carl Sagan's rigid commitment to the rules of evidence that prevail in contemporary science, where supporting data is subject to such brutal scrutiny that, unless one is hardened to the process, can easily be misinterpreted as bullying and discrimination. Science people play hardball. If an advocate and their claim and evidence whither under the heat, it doesn't deserve to be in the game.

In what may be called neo–ancient alienism, there will be new voices, new authorities, and a new kind of literature, and this may all be so unfamiliar and intimidating for traditional ancient alienists that they will long for the "good old days" when their beliefs were comfortably ensconced within the genre known as alternative science. I know that it may be hard—no, *will* be hard—for long-time ancient alien followers to switch, but if they are sincere in wanting to see their beliefs impact the world to the degree that I believe they can, they have no choice but to honor Carl Sagan as the new voice and authority of ancient alienism.

Like Carl Sagan, I am uncomfortable with what I see as a casual disregard of the scientific method, a willingness to bend the rules of evidence, and an unwillingness to engage the legitimate concerns of honest skeptics like Michael Shermer. I believe that at some point and on some level, there needs to be accountability. If one is going to state a claim, one must also be willing to defend that claim in front of the academy. Retreating to the safe haven of alternative science to avoid the sharp barbs of critical scrutiny is not a winning strategy; it's a sign of weakness.

Every year dozens of books are published on ancient alienism, including some by the publisher of this book, that, without exception, follow a prescribed template that takes for granted that the claims and evidence being proposed and advanced are true—claims and evidence that are then rejected wholesale by professional skeptics and mainstream science. Think of what it would be like to have books that advance the basic tenants of ancient alienism from a strictly scientific perspective—literature that challenge the scientific community to openly engage the data, test the evidence, and show where we are wrong.

Those who write books on ancient alienism who make extravagant and unfounded claims will soon realize that there is a new sheriff in town, Carl Sagan, who will demand that self-described and self-appointed ancient alien theorists follow what is known in science as the Sagan Rule, which is that extraordinary claims require extraordinary evidence. Neo–ancient alienism will not tolerate wild, unfounded speculations and flimsy evidence. Conventional ancient alienism theorists take notice: As the Sagan Model becomes better known, there is going to be a major defection of ancient alien enthusiasts who are tired of living on the fringe, who are ready to go mainstream and play by mainstream rules.

I recently traded in my old Samsung 2 for a smartphone. What a difference! Sure, the new technology took some getting used to, but it was worth it. For the millions of devout followers of Erich von Däniken and other old-school advocates of ancient alienism, I strongly recommend that you go for an upgrade, and associate yourselves with Carl Sagan and the model he devised. For millions of others who believe in ancient aliens but are sitting on the fence because they are hesitant to identify with individuals who are routine dismissed as "pseudoscientific," and with literature that is typically classified as paranormal or alternative science, attaching your belief in ancient aliens to Carl Sagan, whose writings can still be found in the popular science section at Barnes & Noble alongside books written by Stephen Hawking, Brian Greene, Freeman Dyson, Richard Dawkins, Neil deGrasse Tyson, and other mainstream scientists, identifies you as someone who understands and appreciates the scientific method and accepts its inherent norms and practices.

Unfortunately, conventional ancient alienism is routinely associated with eccentric iconoclasts who take positions so far outside of the scientific mainstream that they are called weirdos, flakes, and scam artists—individuals that no one in polite company would dare identify with. But state that you believe in ancient aliens because Carl Sagan believed in ancient aliens and you create an entirely new dynamic. Of course, you will be pressed to prove that Sagan was an ancient alien theorist, and that is what this book is about.

Old-school ancient alienism has served its purpose. For a half-century it has kept the flame of truth at least flickering until the Sagan Model was rediscovered and broadly disseminated. Now that there is a science-based model to base our convictions on, it's time for ancient alien enthusiasts to leave the old, adopt the new, and not look back. With Sagan as our teacher and guiding inspiration, neo–ancient alienism is on the cusp of being accepted as mainstream science. To settle for anything less than full recognition of a theory we deeply believe in would be to miss a golden opportunity and to dishonor Sagan's legacy.

There is no reason and no logic for ancient alien traditionalists to reject the Sagan Model of ancient alienism or to turn their back on a famous scientist who was in complete and total agreement with their basic premise. How could any rational person not be in favor of converting what they believe into a more scientifically rigid and intellectually compelling argument? How could anyone who sincerely believes that Earth has been visited by advanced aliens turn their back on new evidence and new arguments that bolster their controversial point of view?

In the coming months and years, conventional ancient alien theorists will be pressed to respond to the model of ancient alienism that Carl Sagan crafted at Stanford University. They will be under pressure to explain how his research stacks up against their own, and how their credentials compare to his. Remember: It was as far back as 1962, years before Erich von Däniken wrote *Chariots of the Gods?*, that Carl Sagan tried to turn ancient alienism into a mainstream scientific endeavor, and it was only because of a vicious NASA conspiracy that his brave voice was silenced and his courageous initiative terminated. Given these proven facts, how can any ancient alien enthusiast not lift up Carl Sagan as the founder and iconic leader of a new and much welcomed iteration of our movement? I urge ancient alien believers to embrace Carl Sagan and his ancient alien teachings, and to elevate their game by adhering to modern scientific standards.

Pop culture ancient alienists are now confronted with a new reality: Carl Sagan, a leading mainstream astronomer, was an ancient alien believer. But that irony, as great as it is, is multiplied many times over with professional skeptics, because Carl Sagan, besides being a mainstream and highly respected scientist, was also one of the founders of modern skepticism. What are Michael Shermer, James "The Amazing" Randi, and thousands of other professional skeptics and skeptic organizations going to do now that the great Carl Sagan, the icon of their profession, has been proven to be a lifelong ancient alien theorist? To this point, they have assumed the ostrich position of sticking their collective heads in the sand and pretending that it isn't really true—when inside their hearts and minds they know that it is. This is an embarrassing response from people who regularly sing the praises of full disclosure and beat the drums for transparency.

Over a span of two years I had tried getting NASA and SETI to openly recognize Carl Sagan's deep and abiding interest in ancient alienism, but they refused to talk to me. For some reason, unknown to me at the time, science-based ancient alienism was a sensitive subject they didn't want to discuss. With the doors at NASA and SETI slammed shut, I turned to professional skeptics, people known for being gregarious, engaging, and more than willing to tackle controversial subjects. Much to my surprise, they responded in the same way. They refused to consider the possibility that Carl Sagan, who, for many, was their mentor and guiding inspiration, was an ancient alien theorist.

It's no exaggeration to say that professional skeptics honor Sagan as a true icon in their industry. His books on the scientific method and critical thinking are classics. One of the world's best-known skeptics, Michael Shermer, has a special place in his heart for Carl Sagan. In the dedication of his bestselling book, *Why People Believe Weird Things,* he writes:

> To the memory of Carl Sagan, 1934–1996 colleague and inspiration whose lecture on "The Burden of Skepticism" ten years ago gave me a beacon when I was intellectually and professionally adrift, and ultimately inspired the birth of the Skeptics Society, Skeptic magazine, and this book, as well as my commitment to skepticism and the liberating possibilities of science.[8]

Because I was from the Sagan school of neo–ancient alienism, and not associated in any way with tabloid models, I had reasonable hopes that skeptics like Michael Shermer and James "The Amazing" Randi, a personal acquaintance of Sagan, would jump at the opportunity to investigate my data to see if it accomplished what I claimed.

I was presenting skeptics with a golden opportunity to correct the popular but errant and misleading notion that Sagan was 100-percent committed to SETI and radio telescopes, when through his own writings he clearly stated that his preferred search method was based on the assumption that aliens have been to Earth in historical times. Sagan insisted that a better strategy than radio telescopes would be to examine ancient manuscripts related to the Sumerians, who built the world's first high civilization. The truth is that Sagan was critical of the SETI experiment from the beginning and openly voiced serious doubts that it would succeed.

I was inviting skeptics to draw a distinction between old-school versions of ancient alienism, represented by Erich von Däniken, and neo-ancient alienism, represented by Carl Sagan's Stanford Paper. It was an opportunity to challenge the scientific credentials of those who call themselves "ancient alien theorists" who, in reality, can't hold a candle to Carl Sagan and his academic credentials and scientific accomplishments. I was hoping that professional skeptics would join me in separating Carl Sagan's intellectually and scientifically robust model of ancient alienism from superficial versions of the theory.

I was shocked and surprised to find that professional skeptics are as adamant in their refusal to discuss Sagan's belief in ancient aliens as NASA and SETI. I quickly realized that my first priority had to be to break through a conspiracy of silence that, even among professional skeptics, has effectively shrouded Sagan's greatest achievement in near total secrecy.

The scientific method is not about perfection at first sight; it's about systematically chipping away at flaws and irregularities until the chaff is gone and all that's left is the pure kernel of truth—and the truth is that Carl Sagan believed that it was aliens who brought us civilization. The paper he wrote in 1962 at Stanford University on ancient alienism was intended to be the scientific foundation on which he would continue to build his career and make his mark on the world. This is an indisputable fact that I am hoping professional skeptics will recognize and publicly affirm.

Professional skeptics perform a valuable service in exposing and debunking bad ideas. Unfortunately, up to this point, the skeptics I have contacted haven't been willing to investigate direct evidence that Carl Sagan was an ancient alien theorist. Without exception they think ancient alienism is a bad idea, and it appears that they don't want the reputation of their hero tainted by having him associated with what they are convinced is pseudoscience.

Despite direct evidence in the public domain that Sagan believed in ancient aliens, it appears that most professional skeptics know Carl Sagan like millions of other people know Carl Sagan: as a tireless ambassador for the scientific method, as a highly effective corporate representative for NASA and SETI, and an accomplished critical thinker. Why do skeptics go to great lengths to attack ancient alienism, and not say a word about science-based research on the subject conducted by one of the world's best-known scientists? Is it possible that they have they been intimidated by NASA? Have skeptics like Michael Shermer, James "The Amazing" Randi, Bill Nye the Science Guy, and many others kowtowed to NASA by refusing to talk or write about the Stanford Paper, or about Sagan's extensive work on ancient alien theory? Have they been complicit in the cover-up?

Engaging Ancient Alienism

NASA and SETI scientists act as if the whole subject of ancient alienism is beneath them—as though it would offend the dignity of their calling to publicly engage an idea that they consider so self-evidently wrong-headed that it isn't worth commenting on. Professional skeptics are just the opposite. They think ancient alienism is so wrong-headed that it is worth commenting on. They rank ancient alienism alongside Bigfoot and UFOs as among their favorite things to debunk. So why aren't skeptics debunking Carl Sagan's model of ancient alienism, or even admitting that it exists?

The truth is that Carl Sagan invested 10 years of his life developing the first and only science-based model of ancient alienism. As one of the founders of modern skepticism, Sagan didn't attack old school advocates of ancient alienism like Erich von Däniken and Zecharia Sitchin for their claim that aliens were on Earth in antiquity, which he agreed with, but for the weak and often-fabricated evidence they submit to support their claims.

Professional skeptics routinely convey the false impression that the scientific establishment is 100-percent against ancient alien theory, but that is not true. In his book, *Lonely Planets,* space scientist and long-time family friend of Carl Sagan, David Grinspoon, writes: "If I was asked, 'Do you believe that the universe is full of extraterrestrial beings, and do you think it possible that some of them are now on Earth, or have been in the past?' I think I'd check the 'yes' box."[9]

Despite NASA's intimidation, there are respected scientists such as Grinspoon and perhaps others who have read Carl Sagan's literature on the subject and agree that past alien visitations to Earth, while not yet a proven fact, are an entirely plausible thesis.

The Solution

Michael Shermer, the chief editor of the magazine *Skeptic,* gives us a possible solution to the problem of professional skeptics refusing to comment on Sagan's Stanford Paper: "It is our job at *Skeptic* to investigate claims to discover if they are bogus, but we do not want to dignify them in the process. The principle we use at *Skeptic* is this: when a fringe group or extraordinary claim has gained widespread public exposure, a proper rebuttal deserves equal public exposure."[10]

According to Shermer, when public interest in Carl Sagan as an ancient alien theorist reaches a certain tipping point, he will presumably respond in kind if he thinks it's not true. In this 20-year commemoration of Carl Sagan's death, I am informing the world about Sagan's belief in ancient alienism—and I will be eagerly waiting for Shermer's response. For me, it seems strange to call Sagan's belief in ancient aliens an extraordinary claim when the evidence is so overwhelming.

In fall 2012, after identifying myself as a fellow secularist and a huge fan who had read all his books, I asked Michael Shermer via e-mail if he would be kind enough to assess evidence linked to Carl Sagan's theory of past alien visitations. It was a very honest and straightforward request, and I innocently (and naively) assumed that because it was connected to a theory espoused by Carl Sagan, whom I knew Shermer held in deep respect and admiration, he would perhaps be willing to assess my data.

The response I got back was shocking. Rather than referencing Carl Sagan's theory of past alien visitations, Shermer tried to associate my data with the ancient alien theories of Erich von Däniken, and then personally attacked me by suggesting that I had to be some kind of religious nut or New Age fanatic, even though I explicitly informed him that I was a secular humanist and 100-percent committed to the scientific method.

To put it mildly, I was stunned. Why this venomous response to such a simple inquiry? Was I missing something? I suspected that something was going on beneath the surface, so I took out a subscription to Shermer's slick publication, *Skeptic,* to see if perhaps my claim and evidence might be addressed in a future volume.

It was, sort of. In 2013, in the Vol. 18, No 4 edition, the front cover featured a cartoonish picture of an alien in a flying saucer levitating a pyramid and, below, another alien wearing a Pharaoh headdress teaching primitive Egyptians how to use a wheel. The caption read: "Did Ancient Aliens Bring Us Civilization? We Examine the Best Evidence."[11]

The cover was such a gross distortion of the historical record that it simply added to my suspicions that Michael Shermer was being purposefully evasive and had no intention of responding to my inquiry in a fair and thoughtful manner. The magazine had the wrong people, the wrong continent, and the wrong millennium! A thousand years before the Egyptians constructed their great pyramids, the Sumerians, in the land between the Tigris and Euphrates rivers, built the world's first high civilization. They were the ones who invented writing, the use of the wheel, and dozens of other technologies and innovations. Without stone, they built the world's first pyramids out of clay with their construction of massive ziggurats.

Even schoolchildren learn that ancient Sumer, not Egypt, is the cradle of human civilization.

An even more egregious error is that Shermer was attacking the wrong ancient alien theorist. If you want the best evidence for an extraordinary claim, one of the rules of professional skepticism, developed by skeptic pioneer Ray Hyman of the University of Oregon[12], is that you investigate the work of its most qualified advocate. When Michael Shermer used the phrase *the best evidence,* he was hinting that there is other evidence related to the question of ancient alienism that is inferior to that of von Däniken and Sitchin, a not-so-subtle dismissal of the evidence I sent to him that supports Carl Sagan's science-based model of ancient alienism that he introduced in his 1962 Stanford Paper.

Let me be clear: Ancient alienism began in the late 1950s and early 1960s, years before Erich von Däniken made his debut as an ancient alien theorist. From that time to the present, the only ancient alien theorist with the academic and scientific credentials to claim such a title was Carl Sagan. It was Sagan who crafted the original theory. Any skeptic who claims to examine the best evidence that aliens brought us civilization, and who doesn't begin with Carl Sagan and his Stanford Paper, is either being purposely disingenuous or is professionally inept.

This is where I find myself in a dilemma. There is no way that I can write about professional skeptics and their position on ancient alienism without offending them. If they are sincerely ignorant of Sagan's 10-year investment in ancient alien theory, they come across as grossly incompetent. On the other hand, if they know about the Stanford Paper, and chapters 32 and 33 in *Intelligent Life in the Universe,* then why aren't they citing Sagan's science-based research on the subject as they go about debunking the tabloid versions of ancient alienism? It smacks of a cover-up.

Professional skeptics cast themselves as Sir Galahads, white knights in shining armor defending the Flame of Truth against the dark forces of ignorance and superstition, and, in most instances, they do a fantastic job. But individuals such as Michael Shermer, James, "The Amazing" Randi, Bill Nye, and hundreds of others professional skeptics may be holding on to a dark secret. They appear to be knowingly suppressing information about the late Carl Sagan that they don't want released to the public, which is that until the day he died, Carl Sagan fervently believed that human civilization was a gift from visiting aliens, and that it was introduced through the Sumerians.

Professional skeptics are in the truth business. Truth is their capital and stock-in-trade. For that reason, their honor and personal integrity as faithful truth-seekers, who are always willing to follow the evidentiary trail regardless of where it may lead, are paramount to their craft. This is why exposing their participation in a massive NASA cover-up involving Carl Sagan, one of the founding fathers of modern critical thought, is one of the most explosive and potentially damaging developments in the history of professional skepticism.

Why are professional skeptics refusing to acknowledge that Carl Sagan was an ancient alien theorist? The most generous explanation I can think of is that they are doing it to protect his reputation as a mainstream scientist. By effectively scrubbing his curriculum vitae of what they consider an unfortunate episode in his life when he was indulging in what many of his peers in the astronomy community called "bad science," they perhaps believe they are doing him a favor.

Even if that were the case, from an ethical perspective it is still terribly wrong. If scientists screw up in major ways, it goes on their record and it can never be expunged. For example, in 1989, two scientists, Stanley Pons and Martin Fleischmann, claimed to have discovered cold fusion. They didn't, and they are regularly mentioned by skeptics in less-than-complimentary terms as examples of credible scientists doing bad science.[13] Another example: In 2011, Professor Antonio Ereditato, a lead scientist at CERN Laboratories, publicly claimed to have discovered a particle that travels faster than the speed of light.[14] He didn't, and he was so thoroughly disgraced that he was forced to resign from his prestigious post. Theoretical scientists are always walking a fine line between fame and infamy. Their calling is not for the faint of heart.

Why should Carl Sagan be treated any different than other theoretical scientists? If he was guilty of a major scientific blunder by proposing in a published scientific paper that aliens have visited Earth, why shouldn't that be listed as a stain on his record, just as it is for other scientists who have disgraced themselves? Why are professional skeptics giving Sagan a pass when no one else gets one? Skeptics like Michael Shermer and James Randi refuse to answer this question, so I will do it for them.

Sagan's belief in ancient aliens was a huge embarrassment to the astronomy community. If he had been a scientist of minor influence he could have been shunted aside with minimal damage to NASA's reputation or to SETI's radio telescope experiment. But Sagan, a public relations genius, was designated to become the face and voice of NASA and SETI. Neither organization could afford to lose him, yet they couldn't allow the public to know that he believed in ancient aliens or that he harbored serious doubts that a radio telescope search would work. At that point, the hard thing to do would have been for officials at NASA and SETI to be honest with the public and address the issue openly and candidly; instead, they did the dishonorable thing and tried to suppress the Stanford Paper and cover up the truth about the man who would go on to become as famous as movie stars and sports heroes. In doing so, NASA stigmatized Sagan's ancient alien research to the point where it can't be discussed without it being demonized by associating it with popular models of ancient alienism that are considered outside the scientific mainstream.

Identifying Carl Sagan as an ancient alien theorist is sure to infuriate NASA and SETI leaders and professional skeptics who routinely use his name and fame in their literature to bolster ideas about ETI that, in truth, he would not have agreed with. Hopefully, some among them will join the cause and help us make Sagan's dream of turning the subject of ancient alienism into a scientifically acceptable field of inquiry a reality.

Professional skeptics who knew about this cover-up should have been exposing this fraud. Instead, they ended up participating in it. It was as if the dog guarding the henhouse had fallen in league with the fox. The reason professional skeptics don't want the public to know that Carl Sagan was a life-long ancient alien theorist is because Sagan was as much the father of modern skepticism as he was the public image of SETI and radio telescopes. Skeptics have created a false choice narrative that states that anything that has to do with ancient alienism is, *a priori*, pseudoscientific, and that any individual who advocates ancient alienism is, by inference, a charlatan. Leaders of all three institutions—NASA, SETI, and professional skepticism— have a vested interest in hiding the truth that Carl Sagan, the great alien-hunter and skeptic *par excellence*, never, throughout his entire life, recanted his view that Earth has been visited by aliens.

By openly stating that Sagan was the father of science-based ancient alienism, some might think I run the risk of being sued by the Sagan family for defamation, or attacked by professional skeptics for disseminating a lie, but neither of these things will happen, because it's a truth that is easily verified. Even Keay Davidson, Sagan's official biographer, makes many references to his belief in interstellar spaceflight and past alien visitations in her book, *Carl Sagan, A Life*.[15]

If professional skeptics ever come to their senses and admit the truth that Carl Sagan was an ancient alien theorist, they will be forced to investigate the Stanford Paper and get up close and personal with such science-based concepts as the Drake Equation, Fermi's Paradox, and Moore's Law as applied to advances in space science and technology, probability theory, logic, the size, age, and evolution of the Universe, exoplanetology, exobiology, the Mediocrity Principle, Sumerology, archeology, anthropology, mythology, ethnology, linguistics, and much more. Not the least of their problems is that they will have to explain how long-lived aliens couldn't have reached Earth when the Pentagon and NASA are currently working on building a starship. Slapping a cartoon of an alien in a spaceship levitating a pyramid on the front cover of a magazine won't cut it.

Are there no skeptics who have read the book *The Scientist as Rebel* by Freeman Dyson, and considered the possibility that professional skepticism, having grown ossified and overly cautious with age, needs its own rebels to challenge the status quo and shake things up? I offer one extract from Dyson's book: "[T]here is no contradiction between a rebellious spirit and an uncompromising pursuit of excellence in a rigorous intellectual discipline. In the history of science, it has often happened that rebellion and professional competence went hand in hand."[16]

Tough Love

I consider myself a true friend of professional skeptics, and I am demonstrating that friendship by imploring people who work in a profession I care about deeply to own up to their wrongdoing, to ask for public forgiveness (which they would get), to learn from their mistake, and to then move on to bigger and better things.

Now that the truth about Carl Sagan being an ancient alien theorist is out in the open, professional skeptics are at a crossroads. If they choose to keep pretending that Sagan wasn't an ancient alien theorist, the hole they are digging for themselves will only get deeper. They can't win this fight because the truth is not on their side. While I'm exposing their cover-up, I'm also extending them a ladder that will enable them, rung-by-rung, to climb out of the deficit of credibility they find themselves in and win back the confidence of the people they serve. Whether skeptics do it individually or collectively, they just need to tell people the truth about Carl Sagan and the Sagan Model of ancient alienism, and the sooner the better.

Any way you choose to look at it, the Sagan Model deserves to be taken seriously as a unique, free-standing theory of ancient alienism that was developed in good faith by one of the leading scientists of the 20th century, a brilliant man who deserves to be recognized by both friend and foe as an expert on the subject. The mistaken claim that Sagan's Stanford Paper was falsified early on because alien interstellar spaceflight is impossible has itself been debunked.

Friends of Carl Sagan have every right to be outraged that professional skeptics have failed to accurately represent what Sagan believed—when the evidence that he believed in ancient aliens is on the record, even posted online. They have a right to demand that Carl Sagan's Stanford Paper be placed on NASA's table for active consideration, and that any evidence that supports that model be professionally adjudicated with full transparency.

At some point in the near future I hope to receive a copy of *Skeptic* magazine and find a picture of Carl Sagan on the cover, along with the caption "The Stanford Paper—the Best Evidence That Aliens Brought Us Civilization?" Some would like to pretend that Sagan merely dabbled with ancient alienism early in his career and then left it just as quickly, but the facts do not support that conclusion. Quite the opposite: Until the day he died, Sagan never reneged on the content of his Stanford Paper. He remained a dedicated ancient alien theorist to the very end, and the paper he was working on when he died had to do with long-lived alien civilizations and their exploration and colonization of the Galaxy.

Michael Shermer and other professional skeptics need to openly concede that Carl Sagan was an ancient alien theorist, of a different kind, and join with me and others in celebrating his courage in standing up to an establishment that still refuses to engage his research on the subject. When they begin to systematically and exhaustively investigate his Stanford Paper as a plausible neo–ancient alien theory that bears little resemblance to the tabloid models, they can educate the public on the scientific method by citing the distinctions between the two.

Carl Sagan: Master Skeptic

Carl Sagan is revered by most professional skeptics as one of the founders of their movement. In 1969, he helped launch the Committee for the Scientific Investigation of Claims of the Paranormal (CSICOP), an organization comprised of leading scientists that was created to combat pseudoscientific ideas that were corrupting young minds and undermining scientific education. It now goes under the more workable name the Committee for Skeptical Inquiry (CSI) and features a popular magazine and a number of Nobel Prize–winning scientists on its advisory board.[17]

In the secular world, professional skeptics serve as the equivalent of priests and ministers, proclaiming truth and exposing error. For that reason, millions of people around the world who embrace secular values expect professional skeptics to carry out the valuable service they render with the highest ethical standards and an unwavering commitment to truth and transparency. This is why it brings me no joy to write a chapter that accuses professional skeptics of colluding with NASA and SETI in a massive, long-term cover-up to keep the public from learning that the late Carl Sagan, throughout his illustrious career, was a staunch believer in past alien visitations to Earth.

While busy attacking and exposing ancient alien claims associated with Erich von Däniken, skeptics were busy building a wall of secrecy around Carl Sagan, the famous astronomer who wrote a NASA-funded research paper six years before von Däniken's book, that argued on scientific and historical grounds that it was aliens, the Apkallu, who brought civilization to humans through the Sumerians, a people of unknown origin who spoke a language that scholars have been unable to trace to any other.

One has to wonder: If professional skeptics are convinced that advanced extraterrestrials have never been to our planet, then why do they attack von Däniken and not Carl Sagan? Is Sagan off limits because he almost single-handedly turned professional skepticism into a respected and, in some cases (such as Michael Shermer), highly lucrative profession?

If that's the case, then skeptics across the board are guilty of a fundamental and inexcusable violation of one the guiding principles taught by Carl Sagan, the master skeptic: In the pursuit of truth, show favoritism to no man, regardless of status or influence. There is no room for sentimentality in science.

In a normal and intellectually healthy environment, competition between scientific hypotheses is considered not only desirable, but is eagerly sought after. The overriding reality is that a terrible injustice was afflicted on Carl Sagan that, on this 20th anniversary of his death, needs to be made right. It is my hope that current NASA director Michael Bolden will agree, and allow the Sagan Model of ancient alienism into its sphere of interest. For NASA to do any less would only be a further desecration of Sagan's legacy.

NASA's New ETI Search Strategy

Sadly, the entire subject of extraterrestrial existence today is so muddled, so bereft of the kind of deep and sustained thought and technical analysis that Sagan put into his Stanford Paper, that we find ourselves lost in the weeds. NASA and SETI claim that long-lived alien civilizations probably exist, but they refuse to entertain the possibility that they have been to Earth. Carl Sagan, meanwhile, was a lifelong ancient alien theorist, but for more than 50 years his work has been covered up. While all this was going on, a recent poll indicated that one in three adults believe it possible that extraterrestrials have been to Earth[18], which explains why fringe science books on ancient alienism have sold in the tens of millions.

How does a rational individual go about finding clarity in such a toxic environment? The first order of business is to blow the whistle on NASA's conspiracy to bury the evidence that proves that Carl Sagan believed in ancient aliens, and the place to begin is to tell the world about the Stanford Paper and *Intelligent Life in the Universe,* the two products that Sagan intended as the foundation for a science-based model of ancient alienism. The reality is that the failure of radio telescope SETI to intercept an alien signal leaves Carl Sagan's ancient alien theory and NASA's current search to find extraterrestrial microbial life as humankind's two best remaining hopes for answering the question: Are we alone?

SETI's deployment of radio telescopes to make contact with extraterrestrials too far away to come to Earth was a top-down approach. It was searching for advanced intelligent life, not invisible microbes. But SETI's failure to intercept an alien signal has forced NASA to change strategy and invest in a bottom-up approach that is looking for exoplanetary mold or slime rather than high intelligence. The idea is that if extraterrestrial slime is found, the discovery of alien intelligence will follow. This completely ignores the commonsense rationale that if advanced extraterrestrials exist, it is all but certain that they would own the technology and have had the time to have reached Earth. So, why isn't that possibility being addressed? Sagan's model of ancient alienism is in NASA archives. NASA administrators know it's there. Unless there is an official cover-up going on, why is it not being activated? Why is it still being ignored?

NASA's search for simple life in space has already enjoyed a significant measure of success. First, its astronomers began looking for other planets, and they found them—a lot of them. It is now scientific fact that planets began forming shortly after the Big Bang and that there are at least 17 billion of them in our Galaxy alone. The next step was to find planets in the Goldilocks zone, the right distance from their host star where they are neither too hot nor too cold to support life, and NASA is finding them in abundance. Next was to find planets or moons that have water, a necessary component of life, and they have been found in our own solar system, so we know that there is a plentitude of water throughout the Universe.

NASA's current mission is to find planets or moons that have atmospheres hospitable to life. After that, it's to find planets or moons that have the essential building blocks of life, such as methane and carbon. Finally, the ultimate goal of NASA is to confirm the existence of life, any life, on a moon, planet, or comet that did not originate here on Earth. This is the gold ring that NASA scientists are hoping to snatch. If microscopic life has evolved independently outside of Earth, it will confirm what is known as the Mediocrity Principle, the concept that there is nothing about Earth that is so unique or special that one can safely conclude that ours is likely the only planet in our galaxy where life has evolved. At a 2015 Washington, DC conference, chief NASA scientist Ellen Stofan daringly predicted that primordial extraterrestrial life would be found in the next 10 to 20 years.[19]

NASA's search for exoplanetary microbial life and the Sagan Model of ancient alienism are both science-based and non-duplicative. In the interests of scientific progress, one would think that NASA would allow the Sagan Model to compete against what I call the Slime Model. Not surprisingly, that has not happened, leaving the Slime Model with no serious competition. This is tragically reminiscent of the extreme bias that was shown in favor of the Drake Model in 1964 when a potential competitor, the Sagan Model, was first rejected. As Yogi might have said, this is like déjà vu all over again.

The demise of the radio telescope experiment has created a vacuum, and an opportunity for NASA to formally launch another strategy to complement its search for exoplanetary microbial life. Carl Sagan's model of ancient alienism would seem to be the perfect candidate to fill that void. Unfortunately, the chances that NASA will resurrect Sagan's thesis and activate an Earth-based search to compete against its Slime Model appear vanishingly small. The engrained bias in the astronomy community against ancient alienism is still alive and well.

If the proponents of a theory have a strategy and that strategy is implemented—and the results falsify the theory—the advocates, if they are honorable, will own it. A good example is the SETI Institute, whose theorists, all credible scientists and scholars, openly admit that, after more than 50 years, they have found no hard evidence of extraterrestrial existence using radio telescopes.

Of course such brutal candor does nothing to help SETI build a following or to raise financial support. Quite the opposite: The use of radio telescopes to find an alien signal is a fast fading experiment that appears destined to go down in scientific history as an abject failure. The harsh reality is that people don't support losers, and SETI has lost.

Time to Change

In 1960, searching for evidence of extraterrestrial intelligence in the electromagnetic spectrum with radio telescopes may have been a long shot, but it wasn't a stupid thing to do. Cocconi, Morrison, and Drake knew we had the technology, so, what the hell, let's give it a try and see what happens. Nothing ventured, nothing gained. Well, we've been at it for more than 50 years and nothing's turned up. Seems kind of strange, doesn't it? So what should we do now? Keep on searching another 50 years in the hope we finally detect something, or convert our radio telescopes over to more useful purposes and shift our resources over to other strategies? As President Barack Obama has said, in reference to the failed United States Cuba policy, "I know in my bones that if something's been done for fifty years and it hasn't worked, it's time to try something different."[20] How about instead of looking for intelligent life, we start searching for signs of any kind of life? NASA thought that was a great idea, and it is now actively engaged in looking for signs of simple life forms on other planets in other solar systems. Good for them.

Here's another idea: Why not look for evidence of extraterrestrial intelligence here on Earth? After all, if advanced aliens exist, they would have had plenty of time to explore the entire Galaxy, and a past alien visit to a planet as attractive as Earth would not seem out of the question. In fact, it seems highly plausible.

Despite NASA and SETI stonewalling, confirming that Carl Sagan was an ancient alien theorist has been easy. The proof is abundant and in his own writings. The big challenge now is to prove whether his amazing hypothesis is true or false. If Sagan was wrong, and Earth has not been visited by aliens, his claim could be abandoned for scientifically legitimate reasons, but that has yet to be established. Sagan's model was abandoned, not after an exhaustive search, but out of military ambition and political expediency. NASA, with its all-out commitment to radio telescopes, didn't want non-space-orientated competition, particularly competition generated by one of its own astronomers. Now that the radio telescope experiment has failed, NASA has pivoted into building an interstellar starship, hoping that no one will ask the question that Carl Sagan asked: If we are within a century or two of embarking on the exploration of interstellar space, shouldn't long-lived extraterrestrials have been at it for millions of years?

Incredibly, the censorship and suppression that Sagan experienced from his fellow NASA astronomers in 1963 continues at NASA, at the SETI Institute, and among professional skeptics. NASA's conspiracy to deligitimize Sagan's research, beginning in 1964 and ending in 1996 when he passed away, hinged solely on its resolute denial of interstellar spaceflight. It was the Big Lie that dashed Sagan's dream of seeing the study of ancient alienism become a mainstream science. For more than 30 years, denial of interstellar spaceflight was the norm at NASA—but, my, how times have changed. In both science and in popular culture, interstellar spaceflight, for both humans and aliens, is now considered inevitable. In a stunning reversal of position, NASA is currently involved in two projects that have the goal of building a working starship by the end of this century. Today, one would be hard-pressed to find a single NASA scientist who would go out on a limb and insist that interstellar spaceflight is a forever impossibility for humans, and certainly not for advanced alien civilizations that may be millions or even billions of years older than our own.

Given these new realities, the Sagan Model that astronomers in the 1960s lampooned as ludicrous because it required interstellar spaceflight for aliens, should have been revisited a decade ago when human interstellar programs got underway. But it wasn't. Incredibly, Sagan's research is still being concealed, and the long-term suppression of his theory of ancient alienism continues to this day.

CHAPTER 6

Sagan Under Siege

*There are none so blind as
those who will not see.*

—John Heywood, 1546

To this point I believe that I have made what I think is a compelling argument that a core component of the Sagan Model of ancient alienism, interstellar spaceflight, is a near-inevitable capability for long-lived alien civilizations. To put the finishing touches on my argument, I suggest that the ancient alien debate can be reduced to two simple premises: (1) If ETI *can't* get here, they couldn't have been here, and (2) if ETI *can* get here, it's a near certainty that they have been here. Over the past 20 years, extraordinary developments have been taking place within NASA that lend significant scientific weight to the Sagan Model of ancient alienism, the most notable being the complete reversal of what for decades was a cornerstone NASA doctrine: that interstellar spaceflight is impossible because overwhelming technological challenges and Einstein's cosmic speed limit meant that aliens could not have physically traversed the vast distances of space to get from their planet to ours.

The negative premise, that interstellar spaceflight is impossible, was a NASA/SETI cornerstone for more than 30 years.[1] That cornerstone has now been wiped away by the stark reality that the United States is currently building an interstellar starship that will be up and operational by the end of this century. Yet, despite the obvious, there are undoubtedly a few stragglers around who still are thinking that star travel is a forever impossibility. If you're one of them, this chapter is for you.

In the early 1990s, NASA's long-time denial of interstellar spaceflight, and the reason it rejected the Sagan Model of ancient alienism, became a major roadblock to Pentagon plans to design and build an interstellar spacecraft by the end of this century. It envisioned a fleet of manned interstellar spacecraft armed with futuristic weapon systems that would keep the United States and the West militarily ahead of their rivals. I know, it sounds like science fiction, something straight out of *Star Wars,* but it's true, and the force behind this ambitious project is the Defense Advanced Research Projects Agency, or DARPA.

Created in 1958, DARPA is the lead agency guiding the United States and the West into a future where interstellar spaceflight will be commonplace. But in 1964, it was just the reverse. Back then, DARPA had a profound influence on NASA's decision to bury the Sagan Model of ancient alienism—on the grounds that interstellar spaceflight was impossible.

An excellent book on DARPA is *The Pentagon's Brain* (2015) by bestselling investigative journalist Annie Jacobsen.[2] In cross-referencing Jacobsen's extensive research against my collection of Sagan biographies, I found an abundance of direct and anecdotal evidence that Carl Sagan, in 1964, was so totally and completely devastated by NASA's violent rejection of his model of ancient alienism that it changed his career and fundamentally altered his outlook on life. His dream of seeing ancient alienism research become a part of the scientific mainstream, years before the subject was turned into a pseudoscientific circus, was shattered. Invisibly, behind the scenes, it is clear that DARPA played a key role in destroying that dream.

Think of the Pentagon as a kind of matryoshka doll. Ensconced within a defense department wrapped in seclusion and secrecy is the even more secret and semi-autonomous advisory unit, DARPA, and within DARPA there is the elitist think tank Jason, named after the mythical hero in *Jason and the Argonauts*. According to Jacobsen, DARPA "is the most powerful and most productive military science agency in the world," that "acts swiftly and with agility, free from standard bureaucracy or red tape."[3] Jacobsen goes on to explain in great detail how the Jason group, comprised of the crème de la crème of Western scientific/military knowhow, visualizes and then actualizes the development of breakthrough weapon systems. She describes it as "one of the most secret and esoteric, most powerful and consequential scientific advisory groups in the history of the U.S. Department of Defense."[4] If DARPA is the Pentagon's brain, Jason is its prefrontal cortex.

In the late 1950s and early 1960s, Carl Sagan was one of a number of young civilian scientists that DARPA and Jason recruited for military related research and consultation, and, though his work remains classified, we can assume that he performed with his typical brilliance and high energy. Holding Secret and Top Secret clearances, Sagan would have been privy to national military secrets involving nuclear weapons, over-the-horizon armament systems, and exotic plans to weaponize space.[5] At the same time that Sagan was on the Pentagon's payroll, he was putting the final touches on his theory that, over geologic time, extraterrestrials have likely visited Earth on thousands of occasions, including, most recently, in the historical era.

NASA was created in 1960, two years after DARPA, presumably to conduct non-military space related research that would enable the United States to send astronauts to the Moon and back. It was to NASA that Carl Sagan turned for the financial assistance that enabled him to finish crafting the Stanford Paper, which he then advanced as a competitive search strategy to the Drake Model that recommended using radio telescopes to intercept alien electromagnetic signals.

As Sagan was soon to find out, NASA is not an independent decision-making body. Like every other government agency that conducts advanced scientific research, it owes primary allegiance to DARPA, Jason, and the Department of Defense. An analogy of what NASA is like would be the Moon: It has a bright side that is plainly visible, and a dark side that can't be seen. Those of us in the public are permitted to see NASA's bright side, with smiling astronauts doing somersaults in zero gravity and stunning photographs of distant stars, constellations, planets, and moons. We are not allowed to look into NASA's dark side, into its secret military related research.

The reality is that behind everything NASA does, including its smiling astronauts and breathtaking photography, is a military component that, according to Jacobsen, ensures that "DARPA technology is ten to twenty years ahead of the technology in the public domain."[6]

In this sense, NASA is no different than "hundreds of research projects—involving tens of thousands of scientists and engineers working inside national laboratories and defense contractor facilities, and university laboratories—all across America and overseas."[7] NASA is simply one of many scientific agencies contracted by the DOD to conduct specialized research that DARPA and Jason use to develop highly classified advanced weapons systems designed to keep the United States and the West decades ahead of their enemies.

In 1964, the Drake Model had the advantage of being militarily useful. It advanced new applications for radio telescopes and involved code identification and decryption, assets that could be mined by the Pentagon. Plus, while monitoring space for alien signals, SETI's radio telescopes could, and did, intercept communications generated by Russian satellites and aircraft. Under the guise of searching for alien radio signals, it had the ability to spy on our terrestrial enemies. In the heat of the Cold War, radio telescope SETI was considered a valued addition to American intelligence services.

Sagan Model Deficiencies

The Sagan Model of ancient alienism had no equivalent assets, nothing that would be of potential benefit to the Pentagon. Still, Sagan was hopeful and confident that NASA would approve and support his search plan. Besides having been funded by NASA and being scientifically robust, Sagan's proposal related directly to the two most important questions any human can ask: Are we alone in the Universe, and, if we have company, have we been visited?

When NASA accepted the Drake Model and rejected his search plan, Carl Sagan had to have been crushed beyond words. Through his involvement with DARPA, he would have immediately known why his proposal to establish an alternative SETI had been turned down. It had nothing to do with scientific incompetence or conceptual flaws. It was for one simple reason: Ancient manuscripts can't be weaponized.

It cannot be emphasized enough that Carl Sagan's singular mission in life was to make the greatest discovery in the history of science—to be the first to prove the existence of advanced extraterrestrials. His entire life and academic preparation were all geared to achieving this one goal, and he was absolutely confident that his ancient alien theory put him on a fast track to success. He knew two things from the beginning: (1) that the radio telescope experiment would end in failure, and (2) that if there were long-lived advanced alien civilizations anywhere in the Milky Way Galaxy, there was no way that at least some of them would not have been to Earth.

During the 1980 American boycott of the Russian Summer Olympics under President Jimmy Carter, there were U.S. athletes who had trained all their lives to compete in that one glorious spectacle, and then, through no fault of their own, were suddenly denied that opportunity for reasons having nothing to do with sport. Multiply their disappointment a thousand times over and you can perhaps begin to understand why, soon after NASA rejected his model for bogus reasons, Sagan cut ties with the military, surrendered his Secret clearances, and became an anti-war activist.[8]

Unfortunately for Sagan, in the intelligence services one cannot surrender a Top Secret clearance, walk away, and pretend that nothing happened. The slate can never be totally wiped clean. It was the height of the Cold War. Washington and the military were rife with foreign spies. The Soviet Union had successfully stolen our nuclear secrets and, in 1957, launched the Sputnik satellite with a rocket that could just as easily have nuked New York or Washington, DC. American fear and paranoia were palpable. Overnight, Sagan went from being one of DARPA's bright young superstars to living in constant danger of being accused of treason and espionage. It was, after all, no secret that bitter and disgruntled scientists were fertile ground for enemy recruitment operatives.

Sagan had been a left-leaning Democrat before this incident, which put him at the opposite end of the political spectrum from the hyper-patriotic, super-conservative Republican types that filled the Pentagon. On top of that, he was the Jewish son of Russian immigrants, and was fond of collaborating with Russian scientists and academics. It was a time when Senator Joseph McCarthy, in his efforts to find and root out the Communists that he imagined were hiding behind every rock, was running roughshod over constitutional privacy rights. It is safe to assume that Sagan was fully aware that he was under constant high surveillance. For the next 30 years of his life, the hotel rooms he stayed in would have been bugged, his phone tapped, his travels monitored, and his speeches and writings carefully analyzed for any sign that he had become a Communist informant.

Think of this personal ordeal as a kind of slow assassination: the constant pressure and stress of knowing that he dare not tell anyone, not his family nor closest friends, about what he knew about DARPA, and being aware that he was being watched. Besides having to constantly look over his shoulder, the bitterness Sagan would have felt against a military-industrial complex that had cavalierly dismissed his most cherished research project as being outside its sphere of interest had to gnaw at his soul. But is there a more sinister element to this story? Is it possible that Sagan contracted cancer after being intentionally exposed to a radioactive substance?

A Timely Death

Sagan's death closely resembled that of fellow American scientist John von Neumann. Both died unexpectedly and relatively young of cancer (von Neumann in 1957 at the age of 54, Sagan in 1996 at 62). The prevailing theory is that von Neumann contracted his cancer after having accidentally ingested a small amount of plutonium at Los Alamos while working on the development of the atomic bomb. DARPA, always quick to seize upon a new opportunity, and with ready access to radioactive material, saw in von Neumann's death a brilliant way to wage stealth warfare on a personal scale without fear of being accused of murder. By the 1990s, clandestine assassination through radioactive exposure had become so perfected that numerous nations had developed the ability to kill in this manner. A recent incident occurred in 2006, when Alexander Litvinenko, a top-tier Russian intelligence officer who had defected to England, died by exposure to radioactive polonium 210 that had been dumped in his tea. Afterward, a British inquiry placed the blame directly on Russian President Vladimir Putin.

Let me be clear: I have no direct evidence that Carl Sagan was assassinated by the Pentagon. But it is well known that one of the things that can cause myelodysplasia, the rare disease he succumbed to, is exposure to radioactivity. Whatever the cause of Sagan's cancer, there is no denying that his death was an extremely fortuitous event for the Pentagon. By 1990, it was done with the radio telescope experiment. It had long before gleaned all the military-related information it could from the Drake Model and was ready to move

on. Frank Drake and SETI director Jill Tarter were notified that if
an alien signal wasn't intercepted by the year 2000, the experiment
would be abandoned. Frank Drake, with obvious anxiety, predicted
that contact would be made by the turn of the millennium.[9] It didn't
happen.

The 1967 Outer Space Treaty that bans the use of nuclear
weapons in space does not prevent the development of non-nuclear
armament systems. Whether the enemy is Russia or China,
DARPA has always been determined to hold the high ground in
the space arms race. At the turn of the century, the next step in
the militarization of space was to colonize Mars, and there was—
and still is—an open question about who will get there first. Will
it be China, an aggressive new challenger for global military superiority, or America and the West? DARPA knew in 1990 that the
technology to reach Mars was already on the shelf. It was just a
matter of convincing the politicians and the public to finance the
venture as a necessary scientific endeavor, and toward that end
the NASA public relations department has shifted into high gear.
The sales pitch to colonize Mars, presumably for peaceful scientific
purposes, continues to build.

But as early as 1990, DARPA was already thinking beyond Mars.
It had its sights set on the next big challenge: to win and hold the
ultimate military high ground by developing interstellar spaceflight
capability. The SETI program was one obstacle to its plan because
it was premised on the view that interstellar spaceflight was impossible. With SETI on its way out, DARPA could put that concern
aside. This left one remaining barrier: Carl Sagan.

The Sagan Model of ancient alienism is premised on interstellar spaceflight, which is presumably why it was rejected in 1964. If
NASA, with DARPA backing, were to all of a sudden begin building
a starship, the grounds for that rejection would be eliminated, leaving Sagan free to once again make his case that Earth is a visited
planet. That is precisely what Sagan had in mind, only this time he

would do it in public. With his charming and persuasive personality, it is highly likely that he would have succeeded in wresting ancient alienism away from the tabloid shysters and have it recognized as an academically and scientifically legitimate field of research. After all, if NASA were to begin building a spacecraft that can go to the stars, Sagan's prediction that long-lived advanced alien civilizations have been exploring the galaxy for thousands or millions of years would almost have to be true. How could it not be? And, if that were true, it would stand to reason that Earth has been visited. As long as Carl Sagan was alive, Pentagon plans to enlist private contractors to begin actively working at building a starship were at a standstill.

In 1994 Sagan was diagnosed with myelodysplasia, a rare blood disorder and a precursor to leukemia. Two years later he was dead, and DARPA immediately began implementing its plan to build a starship. A coincidence?

One thing about scientists is that they tend to be deeply skeptical of coincidences. Was it only a coincidence that SETI's demise, Carl Sagan's death, and the DARPA initiative to build an interstellar starship all happened within a decade? Yes, it's certainly possible, but holding out a healthy suspicion that these three extraordinary events may be related doesn't seem at all farfetched. Unfortunately, the Pentagon is such a locked-down institution that the answers to this uncomfortable question may never be found.

By 1994, Sagan had more than 30 years to build on the model he developed in 1962. What new arguments would he have brought to bear and what new evidence would he have produced that would have added even more weight to his already-imposing Stanford Paper? Equally significant, his was a unique voice in the scientific world, reaching across continents and slicing through ideologies. His platform was all of humanity—the human species. He had the singular ability to sway public opinion on the broadest scale. He didn't have bombs or guns, but he was a master of the public appearance and wielded the power of the pen as few other scientists could do.

In 1990, Sagan was mature, famous, worldly wise, and ready to take his case for science-based ancient alienism to the masses.

Now that NASA, with DARPA backing, is building a starship, the pretense used by NASA in 1964 to reject the Sagan Model—that interstellar spaceflight is impossible—has been exposed as a blatant lie. As early as the late 1950s, famed rocket scientist Wernher von Braun was predicting that man's ultimately destiny would be to explore the stars.[10] Science fiction writers of the time, amazingly reliable precursors of future technology, had long been weaving stories involving interstellar space travel. All of this leaves me to believe that DARPA and NASA leadership knew from the beginning that Carl Sagan was on the mark when he predicted that humans would conquer interstellar space within the next few centuries. NASA's reason for rejecting the Sagan Model of ancient alienism—that interstellar spaceflight is technologically unachievable—was clearly based on lies and deceit.

In chronicling his life, recent Sagan biographers have walked an uncritical and, unfortunately, well-traveled road, buying into a standard narrative that completely ignores Sagan's commitment to ancient alienism. Throughout this book, I have been introducing direct and circumstantial evidence that, in 1964, NASA implemented a plan that was hatched in the Pentagon to suppress the Stanford Paper and cover up the fact that Carl Sagan believed in ancient aliens. It is tragic that Sagan died before he had a chance to tell the world about his belief in ancient aliens and about a search strategy designed to find direct evidence in ancient manuscripts that Earth is a visited planet.

In the early 1990s, Carl Sagan was no longer a young and unknown astronomer. He was the most popular and trusted scientist in the world, and he was ready to break the shackles of silence that for 30 years had kept his ancient alien research in the dark. He knew that he could take his argument and evidence directly to the masses and convince them that advanced aliens, if they exist, could easily have reached Earth. In preparation for his new venture Sagan began writing his second scientific paper on the subject, titled "On the Rarity of Long-Lived Non-Spacefaring Galactic Civilizations." He passed away before its completion.

As the world's best-known scientist and a global celebrity, Sagan could command the attention of international media on a moment's notice. If he were to appear on *Sixty Minutes* or *Today,* and disclose that he was a lifelong ancient alien theorist and that he was certain that extraterrestrials have been to Earth, the world would have gone ballistic with excitement. From New York City to Timbuktu, the news would have made headlines.

After allowing an appropriate time for mourning, and, more important, for people to begin to forget about Sagan's profound influence, NASA, with DARPA backing, quietly launched two competing starship initiatives. One was Icarus Interstellar and the other the 100 Year Starship Project.

Icarus Interstellar and the 100 Year Starship Project

The Icarus Interstellar home page states:

> Icarus—An International Organization Dedicated to Starship Research and Development. Project Icarus is a theoretical design study with the aim of designing a credible interstellar probe that will serve as a concept design for a potential unmanned mission that could be launched before the end of the 21st century. Icarus will utilize fusion based technology which would accelerate the spacecraft to approximately 10% of the speed of light.[11]

Icarus Interstellar director and co-founder Dr. Richard Obousy previously worked for the UK Defense Evaluation and Research Agency; the other co-founder, and president, Dr. Andreas Tziolas, worked at the Johnson Propulsion Laboratory (JPL), a leading

defense contractor. Projects that Icarus is currently working on that
are described on its Website are named Voyager, Helius, Tin Tin,
Forward, XP4, Bitfrost, Hyperion, and Persephone.[12] For a more
comprehensive understanding of Icarus Interstellar, I urge readers
to go to the Website.

With its motto "Exploring the Future of Space Travel," the 100
Year Starship Project "was created by DARPA in partnership with
the NASA Ames Research Center to explore the next generation
technologies needed for long distance manned space travel." It lists
the following five factors as high-level motivations for the exploration of distant space:

- Human survival: ideas related to creating a legacy for the
  human species, backing up the Earth's biosphere, and
  enabling long-term survival in the face of catastrophic disas-
  ters on Earth.

- Contact with other life: find answers to whether there is
  other life in the universe, whether "intelligent" life exists
  elsewhere in the galaxy, and, at a basic level, whether we are
  alone in the universe.

- Evolution of the human species: exploration as a human
  imperative, expansion of human understanding and consciousness through space exploration.

- Scientific discovery: breakthroughs in scientific understanding of the material universe, a pursuit for knowledge.

- Belief and faith: a search for God or the Divine, a need to
  explore beyond Earth's atmosphere as a part of natural the-
  ology or as found through religious revelation.[13]

All of this sounds deeply aspirational, but what it doesn't divulge
is that the core goal of the Starship Project is to develop Star Wars–
type spacecraft and weapon systems capable of knocking the shit

out of any nation crazy enough to take on the United States in an all-out war.

The launch of this initiative was announced in 2010 by retired Brigadier General Peter "Pete" Worden, described as "an expert on space issues, both civil and military."[14]

Starship Project's program manager, Paul Eremenko, previously worked for DARPA's Tactical Technology Office, which was "responsible for drones, robotics, X-planes, and satellite programs. He developed and managed projects to revolutionize design and manufacture of complex military systems (such as vehicles and aircraft) called Adaptive Vehicle Make, the System F6 fractionated spacecraft program, *and the 100 Year Starship*[15] [emphasis added]."

Note that in this bio, the 100 Year Starship Project is specifically designated a "complex military system."[16] This leaves no doubt what the project is about, and it's not altruistic peace-loving science.

True to the Pentagon's strategy of pulling the wool over the eyes of the public to get what it wants, both organizations laud the nobility of conducting groundbreaking space research for the many peaceful benefits that would likely accrue to humankind. Neither site says anything about what they are really about, which is to conquer, control, and defend space by developing new ways to kill and destroy with mind-boggling proficiency.

Carl Sagan knew the truth about the DARPA/NASA relationship—that it's not about white turtle doves and flowers in the hair. Quite the opposite: It's about starships, death rays, and the next generation of stealth technology. I'm convinced that Sagan's goal before he died under mysterious circumstances was to launch a personal campaign to preempt DARPA's plan to weaponize interstellar space. If Sagan had announced to the world that Earth has been visited by extraterrestrials who want our attention and resources focused on world peace and the healing our broken planet, and not on building bigger and more costly weapon systems, I believe that the public would have listened with rapt interest and approval. Had he continued living, Sagan might well have succeeded in under-mining secret military plans to siphon hundreds of billions of tax-payer dollars out of the national treasury, money that it must have to build exotic interstellar weapon systems under the guise of scientific exploration.

In what would have been a major embarrassment for NASA, a high-profile rollout of Sagan's ancient alien theory would have likely incriminated the individuals and organizations responsible for 30 years of suppression and cover-up of his model of ancient alienism. Investigators would have had a field day asking Sagan why his origi-nal research had been rejected. That, in turn, would have created an avalanche of inquiries, formal and informal, that would have led to the doors of the Pentagon. A lot of important people would have been implicated, many of whom were still alive.

The Icarus Interstellar and the 100 Year Starship Project Websites feature dozens of technical papers that are available to anyone, including the Chinese and Russians. Knowing how DARPA operates, it knew when it created these sites that everything on them is information the enemies of the West already have. What is posted is merely the unclassified tip of an enormous iceberg. Beneath the surface and safely out of sight, it is almost certain that DARPA and its secret research facilities have successfully solved the technological challenges of interstellar propulsion systems and have developed the capacity to launch a manned interstellar craft from Mars soon after the red planet is colonized. I believe the DARPA plan is for low-gravity Mars to become a military base, a kind of mothership, for the launching of manned and unmanned interstellar spacecraft. Of course the Chinese know this, which is why they have their own version of DARPA and NASA, and why they have built the world's largest radio telescope—that, in part, they say, is to "monitor" Mars and asteroids as well as to detect possible alien signals. The Chinese are well aware of Pentagon intentions. Largely out of public view, and unknown even to most politicians, the interstellar space race has been in full swing for more than two decades.

To some Pentagon elites, the peace-loving Carl Sagan was a traitor, which is why his name doesn't appear in either the Icarus Interstellar or the 100 Year Starship Website. The question of what caused Carl Sagan's death may be open to debate, but what is not open to debate is that Sagan stood in opposition to long-term DARPA/NASA goals. Though the circumstances surrounding Sagan's death raise suspicions, I don't believe that DARPA, the dark side of the Moon, is inherently evil. The issue I have, and that Sagan had, is that it's morally wrong for an organization like NASA to present itself to Americans and to the world as a benign, science-loving organization that will leave no stone unturned in its quest to find evidence of extraterrestrial existence—and then limit itself only to research that has potential military application. That, I think most would agree, is a fundamental violation of both the spirit and the letter of the scientific enterprise. What does it say about the United States as a civilization when innovative theories and research papers accepted for analysis by our leading scientific organizations are limited to only those that are applicable to death and destruction?

Now that the United States is well on its way to achieving interstellar spaceflight capacity, how can any rational person still say that it's impossible? They can't. The concept of interstellar spaceflight must be scientifically sound and achievable, even for a young emerging species like our own. This means that the Sagan Model, as articulated in the Stanford Paper, is also scientifically sound and achievable.

Isn't it odd that those who argue against interstellar spaceflight never mention Icarus Interstellar or the 100 Year Starship Project? Is it possible that they don't know? Let's give them the benefit of the doubt and say that they have been innocently ignorant. But now that they know, they have no excuse. They can go to the Websites of both organizations and see the facts for themselves. If they have even a scintilla of intelligence and common sense, they have to concede. Unless they are prepared to be labeled as Luddites, those who still deny interstellar spaceflight can't keep living in a make believe world.

But as the exciting new prospect of human interstellar spaceflight in the not-so-distant future sets in, some mending and repair work are to be done concerning the past. The pain and suffering that Carl Sagan endured through most of his adult life as a result of his belief in the inevitability of interstellar spaceflight needs to be acknowledged, and a formal apology by NASA and SETI to the Sagan family is in order. Beyond that, NASA and SETI need to issue apologies to the world for not allowing a scientifically legitimate Earth-based ETI search strategy to compete against the Drake Model, or, for that matter, the von Däniken Model.

Like Sagan, I harbor tremendous concerns about Pentagon intentions concerning outer space and interstellar spaceflight. As long as the military is involved, I see no way for this to end well. Fortunately, it is not too late to do something, but, the fact is, the horse is already out of the barn. It will take a concerted action by people all around the globe to keep these starship initiatives from becoming reality.

SETI now admits that it is highly probable that aliens engage in star travel, just as Sagan surmised. But its long denial of interstellar spaceflight throughout Sagan's 40-year career can't just be swept under the carpet. In my opinion, this was more than a scientific blunder: It was a conspiratorial act that impacted Carl Sagan's hopes and dreams. In this chapter I examine in more detail the tremendous repercussions of NASA's inexcusable violation of Carl Sagan's rights and privileges as a professional scientist.

Frank Drake and other SETI pioneers assumed, in error, that for alien interstellar travel to work, their spacecraft would have to attain velocities that approached the speed of light, and, even then, reaching Earth would take longer than normal life spans. In the judgment of these authorities, this made interstellar spaceflight impossible. Albert Einstein and his cosmic speed limit, they insisted, was the final nail in the coffin of Sagan's ancient alien theory.

They were wrong.

Hindsight is always 20/20, and, in looking back, it seems strange that smart individuals such as Frank Drake were unable to understand that, with billions of years to work with, extraterrestrial civilizations could easily colonize the entire Milky Way Galaxy traveling at velocities even 1/100th of the speed of light. Enrico Fermi, one of history's great mathematicians, was able to crunch the numbers on the back of an envelope. An alien visitation to Earth was such an obvious outcome that when he posed his famous question—Where is everyone?—to a gaggle of fellow scientists, all of them immediately understood and appreciated the gravity of what he was saying.

So how is it possible that it has taken SETI scientists a half century to figure it out?

This appears to be more about conspiracy than credible science. Is it reasonable to think that SETI scientists were so riveted on radio telescopes, so certain that they would work, so thrilled with the anticipation that they would be the ones who would make the greatest scientific discovery in human history, that they just innocently overlooked the common sense logic that Sagan articulated so well in his Stanford Paper? I don't think so.

Fermi understood it, Sagan understood it, Russian astrophysicists understood it. In a 14-billion-year-old Universe, aliens even a mere 10 million years ahead of us would have the entire Galaxy mapped and explored, and every planet that was amenable to life, including ours, visited numerous times. Though it may involve rockets, it's not rocket science.

The stunning thing is that it appears that SETI scientists weren't even willing to work through the math. Without debate, they drew a line in the sand and pronounced with imperialist authority the impossibility of star travel. For the next 50 years this flawed conclusion held the lofty status of an immutable scientific principle. No interstellar travel, ever? It sounded wrong to Sagan 50 years ago and it still sounds wrong today. Like a Papal Bull from the Dark Ages, or a magisterial decree from the Flat Earth Society, the claim that alien interstellar spaceflight is a scientific impossibility begged to be challenged. Sagan was the only American astronomer brave enough to do it, and he paid a heavy price for his audacity.

Challenging the Status Quo

Public esteem for astronomers and rocket scientists in the 1960s was extremely high. Most of us living back then put them on par with clergy and doctors. We considered them so smart that challenging them would never enter our minds. When they spoke, it was with mathematical precision and scientific certainty. At the same time, scientists were equating people who believed in past alien visitations with weird religious cults like Scientology. Not surprising, the public was quick to side with mainstream astronomers and agree that alien star travel was impossible.

But Carl Sagan, a highly talented and supremely confident scientist, had no problem challenging conventional scientific wisdom. In the Stanford Paper, he informed NASA that any reasonably advanced extraterrestrials could get to Earth and that there was compelling historical evidence that they have been to Earth. NASA, the people who held the purse strings, were listening to Sagan in one ear, but in the other ear they were hearing testimony from senior astronomers that interstellar spaceflight was such a ridiculous notion that it wasn't even worth discussing. In what now appears to be an act driven by conspiratorial intent, NASA quietly pulled the plug on the Sagan Model and went exclusively with the Drake Model and radio telescopes.

The Stanford Paper is one of the great documents in scientific history, but, in retrospect, it got Sagan into a lot of trouble.

By writing what he did, when he did, he managed to piss off just about every living individual involved in the space sciences. In vintage Carl Sagan fashion, he blurted out a bold, new theory that was the antithesis of status quo thinking, without bothering to filter his ideas through established scientific channels. Sagan probably knew his theory would be rejected, so why bother? He was apparently hoping that there were individuals within NASA who were independent and unbiased enough to figure out that he was right and that the mainstream was wrong. There may indeed have been some of these kinds of people at NASA, but, in the end, their voices were muzzled and the mainstream conformists won.

Sagan's problem was that what he wrote wasn't bad science, it was bad timing. The Stanford Paper passed a peer review and an editorial review, and anyone who reads it carefully will see that it makes damned good sense. But Sagan, though everyone agreed that he was scientifically brilliant, was at that time young and politically naïve, and terribly impatient with the intolerably long vetting process that is generally the norm in Western science. He had little respect for the system, and—let's face it—in the long run the system usually wins.

What did Sagan write that got everyone so upset? Following is a laundry list of items in the Stanford Paper that challenged conventional thinking. Take your pick.

1. He openly criticized a radio telescope search.
2. He insisted that extraterrestrials have mastered interstellar spaceflight.
3. He posited that aliens have visited Earth in historical times.
4. He gives scientific credibility to the Bible.
5. He collaborated with Russian scientists at the height of the Cold War.
6. By advocating interstellar spaceflight, he fueled the UFO frenzy.

7. He presented a plan to build spaceships that could travel to the stars.

8. He acknowledged the power of Fermi's Paradox.

Any one of these things would have gotten Sagan in trouble, but he managed to put them all into a single package. No wonder he was attacked by fellow scientists as a flake and a kook. Leading figures high in the NASA hierarchy who thought Sagan was crazy for believing in interstellar spaceflight and ancient aliens. Based on that position, what were the chances that NASA would ever change its mind about interstellar spaceflight? It would have taken a miracle, but—guess what—that miracle happened. NASA now not only concedes, albeit ever so quietly, that it's possible that advanced aliens engage in star travel, it has its own interstellar starship development program. Unfortunately, this concession and these developments came too late for Sagan.

Science doesn't tolerate fools and lunatics, but it does occasionally reward innovation and original thinking. Allowing room for new ideas that challenge the status quo is an integral part of how modern science works. Otherwise it risks becoming the very thing that it claims to oppose: a mono-cultural, iron-fisted enterprise that isn't willing to compete against visionaries who dare challenge conventional thinking. Rejecting a credible search strategy because it too closely resembles something going on in the world of pseudoscience is to politicize the scientific method, and that is never a good thing. How NASA treated Carl Sagan and his model of ancient alienism, with so much disrespect, is the low point in its 60-plus-year history.

But Sagan had another problem to deal with: He was a professional astronomer, and what he was proposing in his Stanford Paper was not an astronomy solution. It had more to do with archeology, ethnology, anthropology, history, linguistics, and even religion. Unfortunately, there was no one in any of those softer disciplines in the least bit interested in proving that extraterrestrials were on Earth interacting with the Sumerians, helping them build the world's first civilization.

NASA and SETI East held the strategic high ground on this one. They insisted that the search for extraterrestrial intelligence had to be led by astronomers and needed to be by radio telescopes only. If archeologists wanted to look for an alien signal in ancient manuscripts (and none did), so be it. But for an up-and-coming astronomer like Carl Sagan to espouse a theory that was so completely outside his chosen discipline was intolerable. Sagan may have been the beneficiary of a cross-disciplinary education, but he would have to make up his mind: Choose one or the other. He could opt for radio telescopes and no interstellar spaceflight, and remain a respected member of the astronomy community, or he could believe in ancient aliens, study ancient manuscripts, and be booted out. But under no circumstances would he be allowed to do both.

Sagan finally relented, but never 100 percent. In the face of what he knew would be fierce and sustained opposition from his peers within the fledgling SETI enterprise, he persisted in advocating for an alternative search strategy that was premised on interstellar travel and past alien visitations—and if he had not been such a dynamic and influential personality, his stubbornness would likely have cost him his career.

Thanks to international media attention, SETI and the radio telescope search got off to a fast start. But to sustain the momentum, it needed an articulate and charismatic spokesman, and Sagan was clearly their man. Most SETI scientists were awkward and uncomfortable in front of a camera and a crowd, whereas Sagan, with his good looks and enchanting conversational style, seemed born to be a media star.

SETI and Sagan found themselves at a crossroads. Because of the harsh manner in which the Stanford Paper got rejected, their relationship could easily have turned into a circular firing squad. But in the end, Sagan, under threat of losing his standing in the space sciences community, agreed, under duress, to put his ancient alien theory aside and become a team player. He quickly became SETI's poster child, and he performed with his typical brilliance. Meanwhile, the Stanford Paper quietly drifted off into the dark night of anonymity, where, despite SETI's recent change of mind about interstellar spaceflight, it remains to this day an all-but-forgotten document.

Sagan went on to become the spokesman for NASA and SETI, but, like a mischievous teenager, he always looking for ways to tweak the nose of the establishment. He was constantly and surreptitiously bringing up the possibility of alien interstellar spaceflight whenever he had the opportunity. He quietly advocated for it in almost all of his books, and in his famous *Cosmos* series, he states: "Every star may be a sun to someone. Within a galaxy are stars and worlds and, it may be, a proliferation of living things, and intelligent beings and spacefaring civilizations."[17]

In his bestselling novel, *Contact,* Sagan has both search strategies represented. An alien signal is found using radio telescopes, but the message that was encrypted into the signal is a blueprint for how to build a machine that allows Ellie, the heroine, to travel to the stars using wormholes and time dilation. Sagan's lifelong obsession with interstellar travel had to have driven NASA crazy.

This raises the question: Are there still individuals within NASA and SETI who still think that alien interstellar spaceflight is a ridiculous idea not worth considering? Are there still NASA scientists who openly deny the possibility that extraterrestrials could reach Earth? Are there still people of influence within NASA who would insist that the Sagan Model of ancient alienism should not be revisited under any circumstances? These are disturbing thoughts. Unfortunately, there is evidence of an influential clique within NASA that doesn't want to engage the Stanford Paper or to own up to an ongoing conspiracy against Carl Sagan. They take comfort in assuming that the skeletons of the past are securely tucked away in the closet where they can't be seen. They don't want the door opened for fear the bones might fall out and they will be implicated for being part of a massive and long-running cover-up. This book shines a light on what some at NASA desperately don't want you to know.

Yet, surely unintentionally, NASA has contributed in significant ways to what will hopefully be the eventual formal recognition of Sagan's research on ancient alienism and the acceptance of the Stanford Paper into its hallowed sphere of interest. In recent years, NASA and SETI have made three major contributions to Carl Sagan's ancient alien theory:

1. They have scientifically engaged the deeper implications of Fermi's Paradox.
2. They have officially reversed their positions on alien interstellar spaceflight.
3. They have not formally denied that Carl Sagan was an ancient alien theorist, or denied the existence of the Stanford Paper.

The theory that Carl Sagan developed in 1962 is at last out in the open where it belongs—where it can be critiqued by NASA scientists, SETI theorists, historians, theologians, anthropologists, archeologists and, perhaps most important, by citizen scientists. It's impossible for anyone to appreciate or understand who Carl Sagan was as a human being and as a scientist without being aware of his Stanford Paper, and when, how, and why it was written. The contents of that extraordinary product shaped and molded thoughts and convictions that Sagan would carry throughout his life: bold new ideas that he was about to take public, until his life was snuffed out by a rare disease.

Lest anyone is still thinking that Carl Sagan's belief in interstellar spaceflight and ancient alien visitations were nothing more than the passing fancy of a youthful scientist who, like a defiant teenager, wanted to be different just to be different, I offer the following fascinating exchange between Sagan biographer Keay Davidson and NASA administrator Daniel Goldin, which occurred one year after Sagan's death, as recorded in *Carl Sagan, A Life*. Goldin made a passing comment to Davidson that he was "look[ing] into the feasibility of interstellar robotic missions"[18], an idea he attributed to reading Sagan's *Pale Blue Dot*—but interstellar travel still wasn't being considered by NASA administrators.

A couple of things jump off the pages of these incredible exchanges. One is that in 1997, NASA, through the voice of administrator Dan Goldin, expressed a belief in the possibility of future star travel, and SETI, through scientist Paul Horowitz, called his statement nonsense. For the record, Horowitz lost and Goldin won. Notwithstanding Horowitz's vitriolic reaction, SETI's Website now features the essay on Fermi's Paradox that admits to the plausibility of alien interstellar spaceflight, and NASA, with backing from the Pentagon, is working on building a starship. If, in a century or so, humans will be doing it, we can be sure that aliens have been doing it for a long, long time, which means that there is a high probability that Sagan was right about past alien visitations.

The dots between interstellar spaceflight and ancient alienism are now so close together that they almost touch, but they still need to be officially connected. Who among current mainstream scientists will have the courage to step forward and complete the process? Where are the men and women who will bring redemption and vindication to Carl Sagan?

To illustrate how important it is to the Pentagon to condition the public into accepting interstellar spaceflight as a natural and inevitable step in technological progress, my local small town paper, the *Bend Bulletin*, ran an article from the *Washington Post*, written by Dominic Basulto, about a major breakthrough in quantum computing developed by Google and NASA. Close to being perfected, this quantum computer will be 100 million times faster than any existing computers. In the past, the cited applications for such major advances in computer capability would have been long-range weather forecasting, analyzing global economic trends, and similar complex problems. But in this article, dated December 29, 2015, the potential application was to "optimize the flight trajectories of interstellar space missions."[19] No bold letters, no headlines, just a ho-hum, matter-of-fact allusion to what NASA, for 50 years, had insisted was impossible. Why is NASA so certain that it can safely allude to interstellar spaceflight without raising any eyebrows? Because Carl Sagan is dead.

Exhibit number two: An even more recent article written by Rachel Feltman of the *Washington Post* begins as follows: "If we want to find intelligent aliens with advanced space-faring civilizations, it would make sense for us to start in densest stellar neighborhoods."[20]

"Advanced space-faring civilizations"? This verbiage, which has been lifted right out of the Carl Sagan playbook, would never have appeared in family newspapers just a few years ago. And the author's premise begs to be challenged. If advanced alien civilizations are "space-faring," wouldn't it make more sense to look for evidence that they have been to Earth than to assume that a primitive non-space-faring civilization like ourselves needs to find their home planet to prove they exist?

Coming Clean

NASA desperately needs to hold a major conference on *alien* interstellar spaceflight and come clean on *everything.* SETI's old guard, particularly Frank Drake, needs to answer the question: Are they still absolutely certain that advanced extraterrestrial civilizations have not mastered the ability to physically explore our galaxy—when we have our own starship development program? Yes or no?

Frank Drake and other key leaders in SETI know that this question is like the prisoner's dilemma. If they answer "yes," they are identifying themselves as Luddites and obstructionists who need to resign and get out of the way. Denial of alien interstellar spaceflight is no longer compatible with reality or mainstream scientific thought, which is that the twin barriers of vast distance and Einstein's cosmic speed limit can be mitigated by technological progress and vast amounts of time.

If they think alien interstellar spaceflight is impossible, they need to explain how, in a 14-billion-year-old Universe, spacefaring extraterrestrials could not have evolved billions of years ago—when Earth has produced modern humans in a mere 4.5 billion years.

If they think alien interstellar spaceflight is impossible, they need to explain what, specifically, will prevent the human species from traveling to the stars in search of a new home as our Sun begins to die out and our solar system grows cold and lifeless.

If they think alien interstellar spaceflight is impossible, they need to explain why they have expressed in writing a willingness to test concrete evidence like a landing light that may have fallen off an alien spacecraft, but refuse to test empirical evidence related to an ancient alien theory crafted by Carl Sagan. If interstellar spaceflight is impossible, even for aliens, why not place a blanket moratorium on the testing of all purported evidence of alien visitations?

If they think alien interstellar spaceflight is impossible, they need to explain away Fermi's Paradox to the extent that it can be intellectually and scientifically abandoned. Why, after more than 60 years since it was first uttered, is SETI openly admitting that it's still relevant?

If they think alien interstellar spaceflight is impossible, they need to explain why they are not publicly and aggressively opposing NASA for launching a starship development program. If they are certain that the science is on their side, why are they so quiet and defensive? Why aren't they vigorously trying to purge the space community of individuals like exoplanetary scientist Sara Seager, who has stated, "Could a civilization, if they wanted to, marshal their resources to go to the nearest stars? I think that is within our reach."[21]

Why aren't Frank Drake and others in SETI calling for her ouster? Why isn't she being condemned for scientific heresy? The answer is that belief in the possibility of both alien and human interstellar spaceflight among SETI and NASA scientists has metastasized to the point where it is now the dominant, if not official, position. Those who still deny star travel tend to be old and long past their prime. The young guns at NASA, whether they realize it or not, are siding with Carl Sagan.

SETI and NASA need to hold a summit meeting to clear the air, once and for all, about their position on alien interstellar spaceflight and the theoretical possibility of past alien visitations to Earth. One day we hear from someone at NASA that it's possible, the next day we hear from a SETI scientist that it's impossible. It has to be one or the other, it can't be both. So which is it? Please tell us. I would love to see Sara Seager and Frank Drake openly debate the following question in a public forum: Allowing a few more centuries of continued human development in science and technology, could we humans one day travel to the stars? If this were to happen, I would wager that Seager would win, hands down.

In writings on the subject, I routinely come across estimates that there may be alien civilizations one million years ahead of us in science and technology. In fact, the Universe is old enough and the Earth is young enough that their civilizations could easily be one billion or more years older than ours. Either way, it's impossible for any human to conceive of what such beings might be like or capable of doing.

For the sake of the discussion, let's narrow the gap, by a lot. Let's say that aliens are only one thousand years into the Scientific Age, which, on the cosmic time clock, is comparable to about one minute out of a year. Humans have been in the Scientific Age for 400 years, so, under this scenario, aliens would have a 600-year head start on us in science and technology. The only way to imagine the science and technology that aliens six centuries more advanced than us might have is compare the rate of scientific progress humans have made in four centuries against what we might be capable of accomplishing in the next 600 years. What new knowledge will humans have about the nature and fundamental principles of the Universe in the year 2616? What technological breakthroughs will have been achieved? How many fundamental paradigm shifts will occur between now and then? How many Newtons and Einsteins and Sagans will have been born who will shake the foundations of established science and propel our species in new directions?

Now, let's do something that is extremely insignificant on the cosmic scale. Let's give the extraterrestrials another minute by contemplating what their science and technology might be like if they were 1,600 years ahead of us rather than 600 years ahead of us. What might they accomplish with that additional 1,000 years? Now, let's extend that time to a mere 10,000 years, still 990,000 years short of a million. What would their science and technology be like then?

This is the essence of the Fermi Paradox. It's simply coming to grips with the stunning implications of exponential advances in science that a long-lived alien civilization would be expected to achieve. Why is this fascinating debate not taking place, when it would obviously generate a huge amount of interest and inevitably produce good science?

There can be only one answer: If NASA and SETI openly admit the possibility of human interstellar spaceflight, then it is all but certain that advanced aliens have been to Earth, just as Carl Sagan hypothesized, and that would justify an all-out search for archeological and historical evidence that they were here. That, in turn, would deflect interest and resources away from NASA and divert them to institutions like the University of Chicago and the University of Pennsylvania, both of which have extensive knowledge and experience in the field of Sumerology. Sadly, in the current insular environment, asking NASA to put Sagan's Stanford Paper on the table as an active area of interest would be comparable to asking the board of Coca-Cola to endorse Pepsi. Unless forced by intense and sustained public pressure, it's not likely to happen.

Why are scientists like Frank Drake and David Morrison still hanging around, filling the power positions at SETI instead of retiring to the golf course and letting young blood take over? I think it's because they are haunted by the ghost of Carl Sagan. They now know that they lost and that Sagan won; as they leave, new leadership and bold young scientists will come in and, despite the ongoing conspiracy to cover up Sagan's beliefs and suppress his research, they will embrace the Stanford Paper and Sagan's theory of ancient aliens as being fully compatible with the search for microbial life in the Universe.

How ironic that Dr. Morrison, who is outspoken about his conviction that interstellar spaceflight for both humans and aliens is impossible, is the director of the Carl Sagan Center for the Study of Life in the Universe, a division of SETI. Morrison, one of SETI's elder statesmen, would likely be one who would join SETI chairman emeritus Frank Drake in calling Carl Sagan's Stanford Paper "bad science." They are among those at NASA responsible for keeping the Sagan Model of ancient alienism outside of the NASA perimeter.

The apparent absence of any dedicated thought at SETI on the subject of interstellar spaceflight, human or alien, suggests that it is absolutely impervious to the possibility that aliens have been to Earth. It is simply not on their radar screen, and it should be. In this 20th-year commemoration of Carl Sagan's death, this muddled-headed environment, this division of opinion, is all the more reason for SETI to have an open and honest debate on the subject, all the more reason for NASA to reinstate the Sagan Model of ancient alienism as a theoretically viable possibility, and all the more reason to test evidence that appears to support that model.

In vetting the strength or weakness of competing theories, scientists sometimes find themselves as defendants, standing up for what they believe in, and sometimes as prosecutors, offering critical analysis of alternative models that they think are wrong. Carl Sagan's theory was supposedly rejected on the basis that interstellar spaceflight, even by advanced aliens, was impossible. That position is no longer tenable now that NASA is actively involved in starship construction. It's time for NASA to own up to the conspiracy against Carl Sagan, purge itself of antiquated thinking, and revisit the Sagan Model of ancient alienism.

In his Stanford Paper, Sagan indicated that he didn't think there was a snowball's chance in hell that the radio telescope experiment would work, and he has been proven right. In counterarguments, SETI pioneers like Frank Drake insisted that there wasn't a snowball's chance in hell of either humans or aliens ever developing the ability to physically explore the galaxy. Drake's position now finds itself on shifting sand as more and more scientists are openly challenging that doctrine by voicing the opinion that humans in the not-so-distant future will almost certainly develop the ability for star travel, and, if that's the case, that aliens should have been doing it for a very long time.

CHAPTER 7

Last Words

*He who knows, why should
he keep it hidden?*

—Ancient Sumerian proverb

In December 1996, a physically fading Carl Sagan was admitted into the Fred Hutchinson Cancer Center in Seattle, Washington. There, in his final days and hours, he exchanged endearments with loved ones. His last spoken words, whatever they may have been, are appropriately remembered only by family and close friends. The subject of this chapter is not those words, but, rather, the last eight written words he left behind that are the title of a paper he was working on prior to his death, a scientific essay that he was unable to complete: "On the Rarity of Long-Lived, Non-Spacefaring Galactic Civilizations."

As much as any other evidence presented in this book, these eight words prove beyond any doubt that Carl Sagan was an ancient alien believer and theorist throughout the course of his scientific career. They also prove that up to the very end of his life he remained sharp and in control of his faculties. Suffering through chemotherapy and two bone marrow transplants, Sagan was keeping fully abreast of what was happening within NASA and SETI.

While still holding out hope for a miraculous recovery, he was working on what he knew might be his final paper. It was intended as
a response to what he saw as an existential crisis in the search for
extraterrestrial intelligence. If he were to somehow manage to survive, he knew that there was a golden opportunity to introduce the
world to his unique brand of ancient alienism:

> ". . . interstellar spaceflight is technically feasible—even
> though an exceedingly expensive and difficult undertak-
> ing from our point of view—it will be developed . . ."[1]
> "On the Rarity of Long-Lived, Non-Spacefaring Galactic
> Civilizations"[2]

These two statements, one coming at the beginning of Sagan's
career and the other at the end, are mirror-image observations that
essentially say the same thing from two different perspectives. The
first is a double positive: If aliens exist, they will colonize the galaxy. The second is a double negative: Alien civilizations who don't
explore and colonize the galaxy are extremely rare.

I am convinced that Sagan had his Stanford Paper in mind as he
was working on what he intended to be a farewell statement about
his life and legacy. In true Sagan-esque fashion, Carl Sagan created
a riddle, perhaps wondering how long before people would figure
out that the two statements, like two bookends, not only bracket his
career, but reveal his life-long belief that we are a planet and a species
that has been visited and influenced by a long-lived alien civilization.

In many respects, the title of his paper encapsulates what set
Carl Sagan apart from other ETI theorists. While they were playing checkers, he was playing three-dimensional chess. While they
were fixated on the allure of using radio telescopes to intercept an
alien signal, Sagan was viewing the situation through a 360-degree
panoramic lens. His unique ability to meticulously hone in on fine
details while still seeing the big picture led him to common sense
inferences that rocked the SETI establishment.

His title has four component parts:

Galactic Civilizations: While others were focusing on making contact with a single advanced alien civilization in the Milky Way Galaxy, Sagan knew that if there was one, there were likely to be many. This was not renegade science. Based on the Drake Equation, most ETI theorists agree that our galaxy, with its more than one hundred billion stars, could be populated with thousands of extraterrestrial civilizations, which meant that rather than looking for a single needle in a vast cosmic haystack, they were actually looking for any one of thousands that likely exist.

Long-Lived: Out of a plentitude of alien civilizations, Sagan made the assumption that a statistically significant percentage of them would have survived the difficult childhood and adolescent stages of their evolution and become long-lived. These would be the ones that managed to survive the first precarious centuries of their scientific age, when self-annihilation through nuclear holocaust or the irreversible degradation of their environment would have been most likely to occur. Somehow they managed to figure out the key to long-term survival, knowledge that we humans desperately need to learn before it's too late.

Non-Spacefaring: From the pool of long-lived alien civilizations, Carl Sagan next poses this question: How many of them, for whatever reason, would choose *not* to engage in interstellar space exploration? His use of the double negative was an ingenious strategy that has far more explicatory power and force of logic than asking how many advanced alien civilizations engage in star travel. It's somewhat like asking how many long-lived tribes of primitive humans would *not* have satiated their curiosity and ventured away from their home valleys to explore what was on the other side of the mountains.

Rarity: Sagan answered his own question with his use of the word *rarity.* His conclusion was that the vast majority of long-lived galactic civilization would engage in star travel.

It's clear from the title that the purpose of Sagan's paper was to respond to the growing number of ETI theorists who, by 1990, had finally figured out that Enrico Fermi was right: If advanced alien civilizations exist, they should have made it to our planet, but they aren't here now. To short-sighted SETI theorists, that could only mean one of two things: Either they don't exist, as Fermi concluded, or they do exist but have good reasons why they choose not to make an appearance. Carl Sagan was willing to consider a third possibility: that they exist, have been to Earth in the past, and have since moved on.

Following are brief analyses of the three options:

Option 1: Aliens don't exist. It's a matter of record that many SETI scientists were raised in conservative Christian or Jewish homes and made the heart-wrenching decision to reject the faith of their parents after they were introduced to science. For them, finding hard evidence of extraterrestrial existence, which they fully expected to happen, would be more than a scientific accomplishment. It would be tangible proof that they made the right decision in leaving faith systems that teach that humans are at the center of God's creation.

The thought that humans may be sharing the Universe with other intelligent beings living on other planets has haunted Western theologians and religious leaders for centuries. Confirming the existence of advanced alien civilizations would prove that humans aren't in any way special or unique, and, if we're not special and unique, then neither are the gods that people worship and serve. For the faithful, the possibility that residents of other worlds may worship other gods, or no gods, is fundamentally objectionable.

For evangelical Christians, not finding evidence of ETI would be comforting proof that their literal interpretations of Scripture are correct. The nervous concern within the secular and scientific communities over the SETI failure to make contact has been matched by relief within conservative Jewish and Christian groups who take comfort and reassurance in the possibility that science may confirm that we are indeed alone in the Universe.

NASA and SETI have no one to blame but themselves for this situation. By allowing but a single search strategy rather than encouraging and supporting a robust competition between multiple strategies, they put all their eggs in one basket—and it has turned into a very large and messy omelet. Incredibly, they continue to make the same mistake by endorsing and funding just a single new strategy: the search for exoplanetary microbial life. They still, as of this writing, will not allow the Sagan Model to compete against the Slime Model. For now, the religious right is quietly celebrating SETI's failure, but they know they can't sit smugly. The odds that NASA will succeed in finding microbial life on another planet are extremely high, so high that NASA officials have issued a guarantee that it will happen within the next 10 years.

But there is a potential danger in this approach. The overly optimistic rhetoric coming out of NASA and SETI in the past, promising that an alien radio signal would be intercepted in short order, elevated hope among secularists that victory was assured. Those hopes came crashing to earth when it didn't happen. NASA and SETI may be making the same mistake again with their prediction that the discovery of simple extraterrestrial life is eminent.

But what if it doesn't happen? What if the Slime Model fails? Until the discovery is made, the possibility that the only life in the Universe is what is here on Earth can't be ruled out. If, against all odds, this search ends up failing like the radio telescope experiment failed, biblical fundamentalists would have even more reason to celebrate.

Option 2: Aliens are shy. Not willing to concede that humans are alone in the Universe, and under strict orders from NASA not to revisit the Sagan Model of ancient alienism, SETI scientists had no recourse but to start inventing reasons why aliens were not showing up at their Mountain View, California, headquarters. With one exception, the reasons they came up with were so spurious, even downright silly, that it's almost like they came from a class of kindergartners: "They can't afford it," or "They are so happy where they are at, they aren't interested in exploring the cosmos," or "Earth and humans aren't interesting enough to warrant their attention," or, my favorite, "They watch us from a distance like people watch animals in a zoo." For any of these farfetched scenarios to be true would require that they apply to all alien civilizations in the galaxy, that there be no exceptions. This is so wildly implausible that these speculations were often expressed apologetically, in a tone that said "Sorry, folks, but this is the best we can do." It makes me wonder how many of the ETI theorists coming up with these bizarre ideas were aware of the Sagan Model and were quietly thinking that it would make a lot more sense to give Sagan's search strategy a shot.

Option 3: The Sagan Model. In about 1990, Carl Sagan, taking note of all this foolishness, saw an opening, a golden opportunity to reintroduce his theory of ancient alienism. With typical Sagan subtlety and eloquence, he called non-spacefaring and long-lived galactic civilizations "rare." In his mind, the vast majority of them would be out and about, exploring and colonizing the far reaches of our galaxy, so of course they would have been to Earth, just as Fermi hypothesized. Sagan would have then very discreetly but forcefully suggested that perhaps aliens have been here in the past but have since moved on, and that it might be worth exploring ancient manuscripts that refer to primitive humans coming into contact with godlike beings to see if perhaps they left some kind of encrypted signal or code as proof that they were here.

SETI's failure, the new interest among SETI scientists in Fermi's Paradox, and major advances in rocket propulsion all created a perfect opportunity for Sagan that he couldn't pass up. His common sense and dignified solution left the door open for alien existence, and, at the same time, resolved Fermi's Paradox without resorting to silliness. Given the crisis that the SETI community was in, and the perplexity among the public why contact had not yet been made, Sagan was confident that he could convince both the public and the academy that his model of ancient alienism, that bore no resemblance or relationship whatever to the fluff that was being hawked in the tabloids, deserved serious consideration.

Much of what happens in science is about eliminating and abandoning theories that are proven wrong, and investing time and resources on theories that have a better chance of being true. In 1964, the Sagan Model was rejected without being given a chance to compete. It was eliminated and abandoned on the grounds that interstellar spaceflight was impossible. Though it almost disappeared from the face of the Earth, it's still here, sitting on the sidelines, waiting for an opportunity to enter the game. As theories placed ahead of it failed, there should have come a time when NASA activated a search strategy advanced by one of its most distinguished scientists, but it was clear to Sagan that the Pentagon would never allow that to happen. Sagan, seeing what was going on in the SETI community, seems to have made the decision to take matters into his own hands. He dusted off his ancient alien model with the intent of sharing it with the world, before his untimely death abruptly ended his initiative. For good reasons he was confident that any persons or organizations that might take action to dismiss his initiative, including NASA and the Pentagon, would be so overwhelmed by massive public support that they would have little choice but to get on board.

To secularists battling religionists, SETI's failure revived fears that fundamentalist Christians might be right—that humans are alone in the Universe. It was a major setback, but not enough for them to embrace the Sagan Model.

DARPA, however, saw in SETI's failure a way to turn lemons into lemonade. It decided to embrace the idea that humans are alone in the Universe, and, for that reason, they now argue that it is imperative that humans develop interstellar spaceflight capability so that we can colonize and seed the Milky Way Galaxy with our gifts of intelligence and civilization. Our destiny, we are now being told, is to do what any advanced species would do if it was the only advanced species in existence: go forth and multiply.

DARPA's argument is that our long-term survival depends on humans becoming a spacefaring multi-planetary species. In what are known as black swan scenarios, DARPA scientists remind us that Earth could be struck with a massive meteor that could wipe out life, or that human extinction could come about by nuclear war, or an outbreak of a new airborne influenza resistant to all known antibiotics, or by irreversible environmental degradation. If we were to establish colonies on Goldilocks planets circling other suns, the human species on Earth might be destroyed, but human life in the galaxy would continue on.

All of this makes sense if you don't think about it very deeply. DARPA admits on its interstellar Websites that it wouldn't take many generations for humans living on other planets in radically different environments to become fundamentally different than what humans are on Earth. In other words, to us, they would be aliens; and to them, we would be the aliens. DARPA fails to mention that the aliens we might create would not be under our control or dependent on us in any way. They would have their own governance, their own culture, their own values, and their own aspirations. This might conceivably include invading Earth, wiping out our species, and replacing us with their own kind. After all, they would still possess our darker traits, such as our impulse to conquer, dominate, and destroy. There would be no guarantee that the aliens we created would be kind and benevolent. As a non-human species, they might, in their minds, have good reasons to want to kill the gods that created them.

The saving grace in all this is that despite what DARPA/NASA scientists are saying, the Pentagon has no intention of sending humans to other planets in distant solar systems. It would take too much time and cost too much money. It's all window dressing, a cunning effort to soften public resistance to the colonization of Mars. All the Pentagon is interested in is developing interstellar spacecraft armed with advanced weapon systems that will enable the United States and the West to dominate and control this solar system. The way they see it, the epic struggle between our values and our way of life, and those who oppose us, will be won or lost in stellar space, not interstellar space. If the United States and the West aren't the first to explore, conquer, and colonize all areas within our solar system, we are in eminent peril.

The Galactic Confederation

So far, nothing has been found on any of the planets or moons or asteroids in our solar system that would be worth the effort or expense to mine and transport back to Earth, and whatever scientific research that needs to be done, could be done more safely and more affordably with unmanned rocket ships and robots equipped with artificial intelligence. The only reason to develop interstellar spaceflight capability and to establish colonies on Mars and other places in our solar system would be to maintain and extend military superiority. DARPA's objectives were opposite to Carl Sagan's values, just as his dreams of world peace and global cooperation were regarded as Pollyannaish and hopelessly naive by the Pentagon.

Sagan was right and the Pentagon is wrong on this matter. Long-term survival of the human species can never be won by military dominance. In fact, just the opposite is true: Weaponizing space is the fastest way to human extinction. Sagan saw evidence in ancient manuscripts of a possible Galactic Confederacy of advanced civilizations that owe their long-lived survival to mutual respect and cooperation rather than military domination. He knew that if the human species is to survive and become long-lived, a path must be blazed that will move us in a different direction, away from the Mutually Assured Destruction (MAD) in which we are now headed.

The people who comprise DARPA and Jason are civilian scientists, not military brass. In their minds, what they are doing makes sense, because, while supporting legitimate science, they are also keeping the world safe from tyranny. To them, it's a win-win situation. But they are thinking terrestrially, in many respects not all that different from the anthropocentric thinking of the Catholic Church in pre-Copernican days. The grim reality is that the decision to rush headlong into the weaponization of space may guarantee the destruction of our species, not its long-term survival. My message to the patriotic members of DARPA and Jason is to dare to think about a cosmic reality, the way Carl Sagan did, and consider the possibility that he may have been right about a Galactic Confederation.

To Sagan, this could be the reason why aliens have not made their appearance to modern humans. If they are remotely monitoring the progress of human civilization, they may not think that we qualify to join their club. It may be that the human species must first come together and demonstrate that we have the maturity and wisdom to explore and develop space in unity and in peace. If we can do that, it might allow us a chance to one day become a member of the Confederation. If we can't, then we don't deserve it.[3]

In *Intelligent Life in the Universe,* Sagan references Sumerian legends about the cosmos being governed by a representative and democratic assembly of the gods. Sagan writes: "Such a picture is not altogether different from what we might expect if a network of confederated civilizations interlaced the Galaxy."[4] He also writes: "If interstellar spaceflight by advanced technical civilizations is commonplace, we may expect an emissary, perhaps in the next several hundred years. Hopefully, there will then be a thriving terrestrial civilization to greet the visitors from the far distant stars."[5] Sagan spent 40 years thinking about extraterrestrials in this broad, comprehensive way, which is why, when he died, the hawks in the Pentagon who wanted to weaponized space had to have breathed a collective sigh of relief. In his final years they had to have been terrified that Sagan, the most beloved and trusted scientist on the planet, might break his coerced silence and announce his belief in ancient alienism to the world. The title of his last paper was a defiant shout-out to whoever might be listening about what he stood for and what he believed in. Included in the obituary of Carl Sagan should be a declaration that he died a martyr for universal peace.

Plausible Deniability

By 2010, with Carl Sagan gone, the Pentagon had growing confidence that it finally had an unimpeded path to its development of an interstellar spacecraft, a trajectory that would financially and politically carry it through to the end of the century and beyond. With international terrorism on the rise, the nation was leaning right, toward a more hawkish mindset. With a Republican Congress and hope for a new Republican president, they envisioned generous support from politicians who were alarmed that the United States might be losing its technical superiority in space to the Chinese. They were also confident that there would be broad support from an American electorate nostalgically yearning for a return to the good old days of great accomplishments in space, like the *Apollo* missions to the Moon.

At the same time, the public was beginning to forget about Carl Sagan. Most millennials don't know who he was, and most of those who do only remember his famous Cosmos television series and his support for the radio telescope experiment. In practical terms, the

Stanford Paper was a lost document and Sagan's writings on ancient alienism in *Intelligent Life in the Universe* never existed. On top of all this, tabloid ancient alienism had become so rooted in popular culture that the Pentagon was certain that anyone crazy enough to advocate for a science-based study of the subject would be laughed out of town.

Free from all resistance and confident that it could fend off all challenges, DARPA began to invest in interstellar research. Early in the new millennium, NASA, without fanfare, quietly launched two initiatives though designated surrogates: the 100 Year Starship Project and Icarus Interstellar. Presumably, the goal was to colonize Mars before the middle of the century, and then build a starship capable of sending humans into interstellar space by the end of the century. In the greatest Ponzi scheme of all time, NASA is currently providing cover for the Pentagon's plans to weaponize space by searching for a particular planet in a particular star system that is capable of sustaining human life. We are being told that when humans arrive on that planet and set up camp, our species will have secured a long-term life insurance policy against our extinction here on Earth. All of this is a front to keep the public from knowing the truth.

The question of whether or not there is other intelligent life in the Universe has been successfully subsumed. People don't talk about it much anymore. The goal is no longer about searching for intelligent life, it's to find simple life on another planet that has evolved independent of life on Earth. The high expectation of the public in past years to make contact with ET have been effectively muted. Instead of finding an alien signal, all the talking heads at NASA are now speaking and writing about what a thrilling and historic moment it would be to discover a distant planet that has living extremophiles on it. Though this shifting of priorities from the Drake Model to the Slime Model has left many ETI purists disappointed, their consolation prize is that if NASA finds mold spores on a distant planet in the Goldilocks zone, it is almost certain that somewhere else in the Galaxy there is intelligent life. Just be patient and go with the program, we are told. The search for simple life on other planets initiative, although it may be of some legitimate scientific value, is a smokescreen to cover up the secret intentions of the Pentagon.

The nexus between looking for habitable planets and looking for simple life on other planets is that, at first glance, neither has anything to do with the military. By focusing on these two grand objectives, NASA presents itself to the public as a peace-loving organization dedicated to pure science and the long-term survival of the human species. All reasonable people would agree that both are noble pursuits. But the harsh reality is that NASA is a puppet, and the puppet master, the Pentagon, has zero interest in habitable worlds outside of our own solar system, or in finding simple life on distant planets or, for that matter, in extraterrestrials. The dark side of the Moon cares only about one thing: military superiority by winning and holding the high ground.

This raises the issue: If NASA and SETI were to scientifically confirm the reception of an alien signal and message, would they share it with the public? There is ample testimony from within their own respective organizations that they might not. Following are two examples.

"Humans concerned about their personal and institutional interests might resist the dissemination of some alien information, or seek to brand it as dangerous, immoral, or subversive."[6]

Physicist Paul Davies, chairman the SETI's Post-Detection Taskgroup, echoed Michaud's concerns. He writes:

The first decision would be whom to tell and how. In this scenario, the published Protocol would almost certainly break down. I personally feel that the implications of receiving such a message would be so startling and so disruptive that, although eventual disclosure is essential, every effort should be made to delay a public announcement until a thorough evaluation of the content has been conducted, and the full consequences of releasing the news carefully assessed in light of the Taskgroup's recommendations.[7]

Leapfrog

Like a doe-eyed orphan sitting alone and forgotten alongside a busy highway, the Sagan Model of ancient alienism sits alone on the sidelines of scientific theoretical thought, while more astronomers, cosmologists, physicists, and astrobiologists than one can name or count, in search of a deeper meaning and purpose in the Universe, have been migrating to Eastern mysticism and New Ageism. I am referring to highly credentialed scientists and intellectuals, well trained in Western experimental science, who have thrown caution to the wind as they rush head-long into such esoteric and metaphysical concepts as cosmic consciousness, universal mind, quantum weirdness, parallel universes, multi-verses, alternate reality, and on and on. The scientific imperative to choose the simpler over the more complex would mean that the more reasoned approach would be to consider ancient alienism the better option. Why the mass exodus away from empirical Western science? I believe that one factor is that the great promise of the Drake Model, a thoroughly Western construct, has failed. SETI's ultimate goal was to make radio contact with advanced aliens who would teach us about ultimate Reality. Had it succeeded, everyone would have hailed it as a triumph for Western science, and the exodus to the East would never have happened. Expectations of its success were so high that when contact failed to occur, the pendulum swung to the other extreme, away from Western science and into Eastern mysticism.

Seeing little hope of finding answers to life's deepest questions in the West, scientists began trying to synthesize Western physics with Eastern metaphysics. Fueling this flight to the East has been the Templeton Foundation, a philanthropic organization that issues generous financial grants and awards to Western scientists who come up with innovative new ways of unifying Western science with Eastern religion.[8]

The result has been a plethora of exotic non-verifiable schemes that are so lacking in Western rigor and discipline that 14th-century English Franciscan monk William of Ockham, who invented the law of parsimony, must surely be rolling over in his grave. Also known as Occam's razor, this foundational principle of Western science dictates that when there are multiple explanations for an observed phenomenon, one should choose the simplest, the most natural, and the one requiring the fewest special pleadings.

Hybrid East/West models of cosmology violate Occam's razor with impunity. So, one has to ask: What's going on? Why is it that highly skilled men and women trained in the West have turned to the East, where there little knowledge or appreciation of the value of critical thinking, testing and verification, peer review, quantification, replication, and other principles that serve as the foundation of Western scientific thought? Equally alarming, books on the contorted schemes that flow out of this syncretism are regularly passed off as scientific and appear in the popular science sections of local bookstores instead of in the New Age section where they rightfully belong.

Carl Sagan was an empiricist, and totally committed to Western science. He did not believe in taking shortcuts to the truth by appealing to the voodoo science of either the West or the East, where unseen and unknown forces define and control the fabric of the Universe. Sagan was convinced that the scientific method, though sometimes slow and arduous, was not only the best way, but the only way, for humanity to progress. There is, after all, a good reason why the Age of Science was spawned in the West during the English Enlightenment, and why young aspiring scientists in the East still come to the West to be educated.

The Sagan Model is a thoroughly Western construct, and I maintain that it accomplishes what Eastern and New Age models of reality promise but fail to deliver on. Anyone searching for answers to deep cosmological questions, who doesn't want to abandon the core principles upon which Western science and philosophy have been built, should support the Sagan Model. Rather than being orphaned and forgotten, it deserves a chance to either succeed or fail. Why have so many practitioners of Western theoretical science leapfrogged from the Drake Model into Buddhism without giving the Sagan Model a chance to compete? Because NASA and the Pentagon won't allow it. The Sagan Model has never been formally recognized and activated by the scientific establishment, and just as Sagan was making plans to take an end run around the establishment, his life was taken from him.

A Legacy Moment

Knowing that the odds that he would beat his rare form of cancer were not good, Carl Sagan chose the subject of what he suspected would be his last scientific endeavor with deep forethought. With a career that spanned 40 years, the range of material that he could have written about was literally as large as the Universe. I venture to say that few of his millions of admirers would have guessed that his swan song, his last tome, would be about long-lived extraterrestrial civilizations exploring and colonizing interstellar space, keeping an eye on and occasionally intervening with emergent species like our own.

Though the contents of Sagan's unfinished essay have not been made easily available, it's obvious from the title how Sagan wanted to be remembered. Not as a science popularizer, and definitely not as an advocate for a radio telescope search for an electromagnetic alien signal. Despite being under a strict gag order not to divulge his personal opinion on the subject, Carl Sagan made it crystal clear from this title that he wanted to be remembered as an ancient alien theorist. It is therefore incumbent on those of us who want to preserve his legacy to do everything within out power to make sure his last wish as a scientist be fulfilled, even if it is at odds with the popular narrative being propagated on this, the 20th anniversary of his death.

For 30 years, the NASA and SETI establishments vigorously denied the possibility of interstellar spaceflight. The sole reason for that denial was to ensure that the Sagan Model of ancient alienism would never be taken seriously. Now that NASA and the Pentagon are actively building an interstellar spacecraft, how would it be possible for anyone to argue that Sagan doesn't deserve the highest reward for daring to challenge an established belief?

Unfortunately, Sagan has not been afforded the honors due him because he inserted his research on interstellar spaceflight into a paper that predicted that long-lived aliens have been traversing the galaxy for thousands of years and visited Earth in past ages. The harsh reality is that no mainstream scholar or scientist was willing, or is willing, to seriously and openly consider Sagan's theory, even though, if independently confirmed by new discovery evidence, it would be among the greatest scientific achievements in history.

The deeper reason why scientists and intellectuals are unwilling to take on Sagan's science-based model of ancient alienism has little or nothing to do with science and everything to do with psychology. If science confirms that advanced extraterrestrials have been to Earth and interacted with primitive humans, the Western scientific mindset that has evolved over the past four centuries would be utterly and irrevocably shattered beyond recognition. Like returning to pre-school, highly educated people with advanced degrees behind their names would have to start all over again, from scratch. The Sagan Model would become the Creed for the New Millennium, and many of the most hallowed beliefs of the mainstream establishment would need to be radically reconstructed or be allowed to fall by the wayside.

Scientists estimate that there are about one billion Earth-like plants in our Milky Way Galaxy and that there is a high probability that some of them have intelligent life. Carl's Stanford Paper is a theoretical framework for arguing that it's more probable than not that some of those advanced extraterrestrials have visited Earth. The historical data that Sagan produced in *Intelligent Life in the Universe* is concrete evidence that supports his theory. There is enough credible material to begin an intelligent, science-based conversation on the possibility that Earth is a visited planet. I suggest that it's time that rational people begin that conversation and see where it leads. If Carl Sagan was right, we may now be closer to proving the existence of extraterrestrials than at any time in SETI history, and closer to securing the legacy of Dr. Carl Edward Sagan as the man who led science through a wasteland of false leads and dead-end methodologies to what may at long last be the Promised Land of Contact between ETI and humans.

The reality is that NASA, SETI, and professional skeptics feared Carl Sagan as they feared no one else. Others they could take head on, but with Sagan they had to run and hide. They could spread the sensational claims of others out under the light and show to the world where they were wrong. With Sagan's research, all they could do was to try to conceal and suppress it so that it never found its way into the public consciousness. Now that the Sagan Model is out in the open and in the sunlight, how will they react? What will they do with it? And how will they react to my claim that Carl Sagan died just as he was preparing to display his research openly, in the sunlight, so that it could be properly vetted with full transparency?

Carl Sagan thought that there are likely thousands of advanced alien civilizations in the Milky Way Galaxy that are in constant contact with one another. He also speculated that our species may be so backward and warlike that it hasn't yet earned the right to become a part of the Galactic Consortium. Looking around and seeing our violent tendencies, our gross neglect of basic human rights, our strange anti-scientific superstitions, and our shocking disregard for our fragile environment, Carl Sagan surmised that aliens might conclude that we humans still have a long way to go before we are invited to join the club. If Sagan was right and we have been visited by godlike aliens, then perhaps there is an encrypted key in some ancient Sumerian manuscript that, if found, would unlock a much larger message—a message that we need to help put us put on a course that will propel our species toward global peace and prosperity. If we can move in that direction, even if slowly and incrementally, then perhaps someday we will be welcomed into the presence of our cosmic neighbors.

Humans have been in the age of science for 400 years. If we foolishly manage to destroy ourselves in the next 600 years, we will have had the dubious distinction of joining an unknown number of civilizations that are classified by SETI theorists as short-lived. On the positive side, if we survive as a species for another 600 years, there is a fighting chance that we will become one of the long-lived societies in the galaxy that last for millions or even billions of years. In this hopeful scenario, it is certain that we will have solved many of the major problems we currently face here on Earth, enabling us to venture into deep space out of strength, not out of desperation. Unfortunately, there is a Sword of Damocles hanging over us, and it's in the bloodied hands of a Pentagon war machine that is intent on expanding and escalating the arms race into deep space.

I'm convinced that, in the long run, Sagan will be remembered not for being a science popularizer, but for the model of ancient alienism that he developed in 1962 at Stanford University. Carl Sagan dared to think outside of the box—even while that box was still under construction. His deep commitment to, and lifetime interest in, ancient alienism can no longer be questioned. I am honored and humbled to be playing a very minor role in drawing the world's attention to a sadly neglected aspect of his illustrious career by highlighting his Stanford Paper, which I regard without question as the most brilliantly conceived document in SETI history.

Triumph or Tragedy?

Now that Sagan's work is known, the challenge will be to carve out a place in academia, be it ever so humble and precarious, for the scientific study of the ancient alien model he created. To this end, serious fact-based papers, both in favor of and against his model, will be received and posted on my Website, as will comments and commentary from the general public. The goal is to start building an archive of intellectually honest research material from which current and future generations can learn and to which they can contribute.

Of major concern is what writers of popular and academic books and articles do with the name of Carl Sagan now that he has been identified as an ancient alien theorist. Sagan is, by far, the most cited name in ETI literature, lending instant weight and authority to the subject. Will authors and speakers continue to reference the man who courageously stood up to the military-industrial complex, or will they cave in to the establishment and blacklist Sagan as a scientific leper with whom they want nothing to do? Will future books and articles on ETI distinguish between science-based ancient alienism and tabloid models, or will they continue to perpetuate the lie that there is only one kind—the pseudoscientific? As writers try to figure out what to do with Carl Sagan and his belief in ancient aliens, my hope is that they will take the high road and not allow the work of one of the great thinkers in modern science to fall by the wayside.

Throughout this book I have been hard, but I think fair, on professional skeptics such as Michael Shermer, who, I believe, will play a major, and perhaps decisive, role in what happens to the legacy of Carl Sagan. If skeptics abandon Sagan because of his belief in ancient aliens, it will be a far more difficult challenge to correct and rebuild his reputation as a leader in critical thinking and ETI research. On the other hand, should they embrace the Sagan Model as a scientifically legitimate thesis, it would be a valuable endorsement that would make it easier for other writers to continue to use the name of Carl Sagan in their citations.

Finally, a word of advice to those who may read this book and develop a keen interest in ancient alienism: Don't drink the Kool-Aid that popular ancient alien theorists are serving up. Consuming their gibberish will not only lower your IQ, it will move you in the opposite direction from where Carl Sagan wanted the world to go. It may take a while for the dust to settle, but be patient and wait for credible scholars to weigh in on the subject. For those who have a hunger and curiosity to learn more about the Sumerians and the Sagan Model, and to keep up with unfolding developments, I refer you to my Website.

I am well aware that in this book I have raised more questions than I have provided answers. I hope to resolve that imbalance in the future by sharing additional research material collected over the past four decades, and I look forward to what others may contribute to the cause. In the months and years ahead, what I will be looking for from critics is their considered rebuttal to my claim that Carl Sagan was an ancient alien theorist.

I have hopefully allayed people's fears about visiting aliens by assuring everyone that they are no longer physically living among us, and, while they were here, they were benevolent and altruistic, always acting in our best interests. We can be confident that there is no cause for alarm, and new reason to hope for a bright, exciting, and enduring future for our species. As Sagan often reminded us, we have only just begun our journey to the stars. The next step along that journey belongs to science historians who, it is hoped, will amend the biography of Carl Sagan so that it reflects his deep lifelong commitment to ancient alienism and, in the process, create a new narrative and a restored legacy.

What we do need to fear are the secret collaborations going on among the Pentagon, defense contractors, and high-tech companies such as Google and IBM. There is no doubt in my mind that exotic new weapons are being developed and covert plans are being made in private that will determine the fate of the world—without the world having a voice. If there is one thing we need to fear more than the ignorance and superstition of the masses, it's the immeasurable wealth and unbridled power that are in the hands of a relatively small handful of mostly unknown and unaccountable kingmakers.

In this 20-year commemoration of Sagan's death, those of us who stand in awe and admiration of his life and work have the opportunity to correct an official record that is factually wrong and disturbingly misleading. We need to band together and demand that Sagan be recognized first and foremost as an ancient alien theorist who, in 1962, at Stanford University, crafted a science-based model that predicted that aliens, the Apkallu, were on Earth in the Ancient Near East, teaching principles of civilization to the Sumerians. We need to band together and demand that science analyze and investigate the Stanford Paper in light of recent developments and new discovery data. The true legacy of Carl Sagan is just beginning to come to light, and if Sagan's fans and admirers stand firm, his legacy will survive for the ages.

Boundary Work

In *Preparing for Contact,* author George Michael describes the process that scientist use to separate legitimate science from pseudoscience as "boundary work," which he defines as "to make positive attributions about one's research while denigrating the work of pseudo scientists who ignore or misapply the rules of science." The purpose of boundary work is "to gain standing with peers, granting agencies, and the public."[9]

This is the process that established radio telegraph SETI as a mainstream scientific enterprise. Now that Carl Sagan's work in ancient alienism has been identified and isolated from non-scientific models, will there be NASA and SETI scientists, academics and experts in other fields, and professional skeptics who will step forward and do the boundary work that will be required to make science-based ancient alienism a respected and mainstream area of endeavor?

An indelible benchmark has been established: that Carl Sagan, while still young and in his intellectual prime, became a serious ancient alien theorist who predicted that Earth has been visited by extraterrestrials in historical times. From this point in time and into the future, it is hoped that NASA, SETI, and professional skeptics will openly and without apology add this biographical fact to his extraordinary list of accomplishments, citing it as a badge of honor rather than an embarrassing stain on a truly remarkable and distinguished career. Until this goal is attained, it will be next to impossible to find credible scholars and scientists willing to openly examine the Stanford Paper and other potential evidence that supports the Sagan Model. I and many others will be watching carefully to see what individuals and institutions do with the public disclosure that Sagan was a lifelong believer in past alien visitations to Earth. Will the cover-up go on and the facts continue to be ignored, or will honest truth-tellers step forward and help set the record straight?

This is the fact: Through his Stanford Paper and his book *Intelligent Life in the Universe,* Carl Sagan established a scientific baseline for ancient alien research. His hope and expectation were that he and others would build on that foundation so that, year-by-year, and decade-by-decade, his theory would continue to mature, as new arguments in favor of his thesis accrued, as new supporting evidence was discovered, and as flaws in the model were eliminated. Over time, he anticipated a growing corpus of material related to ancient alienism that would attract more support, fuel more research, and incite more debate.

Instead, what Sagan got from his peers in the astronomy community, who viewed the search for extraterrestrial existence with radio telescopes as a long-term job-security program, were immediate indignation, rejection, and suppression, which, in turn, created the vacuum that pseudoscientific ancient alien theorists moved into with alarming ease and effectiveness.

Carl Sagan never had any desire to "own" the ancient alien genre. He was hoping, I'm sure, to draw bright young scholars and scientists into the discussion who, like him, would not be afraid to challenge the establishment. These would be young visionaries who would generate scientific papers, write books, and hold conferences on Sagan's wildly exciting theory.

Whether young scientists seize the opportunity or not will shed a telling light on the current scientific environment. Is there still so much fear of the Pentagon within the space sciences community that no one will dare to actively and openly engage the Sagan Model? Will the pervasive influence of the military-industrial establishment continue to intimidate would-be enthusiasts who might fear that they would be placing their careers in jeopardy if they were to get involved? Are there old-guard NASA and SETI people around who still think that interstellar spaceflight is impossible, and who will discourage young scientists from contemplating the possibility that Sagan was right?

Sagan wrote that every legitimate science has its pseudoscientific counterpart. What he failed to mention was the order of their appearance. Sometimes it is the pseudoscience that comes first, the best example being young-Earth creationism being widely believed before Charles Darwin gave us the theory of evolution. In other instances, it is the science that comes first and the pseudoscience afterward. A great example of this is quantum mechanics, which inspired, and continues to inspire, a raft of pseudoscientific spinoffs.

Chronologically, Sagan's science-based ancient alien theory came before Erich von Däniken's pseudoscientific model. The problem is that Sagan's work was so effectively censored and suppressed by NASA that von Däniken is widely, and falsely, credited with being the father of ancient alienism. This improper attribution immediately turned the entire genre into a caustic cesspool with such a heavy stigma attached to it that no respected scientist in his or her right mind would dare broach the subject as a plausible possibility.

It is not too late to set the record straight and restore science-based ancient alienism to its rightful place as an academically legitimate area of research. The only way this can be accomplished is to spread the news about Carl Sagan's work on the subject. Because it is clear that NASA, SETI, and professional skeptics aren't interested in making that happen, it will be up to individuals who have been positively impacted by Sagan's life to get the job done. Would you be willing to help? On my Website is a petition that demands that NASA activate the Sagan Model and test all empirical evidence that appears to support that model. Your signature would be appreciated.

Wild and Free

In Northern Idaho there is a river that for thousands of years flowed wild and free, from its snow-fed tributaries high in the Grand Teton Mountains to the Snake River, then to the Columbia, and, finally, to the sea.

In the late 1960s and early 1970s, a massive government agency, the Bureau of Reclamation, decided that the Teton River needed a dam, despite the fact that every environmental, economic, and geological study advised against it. More than 300 feet high and 1,700 feet across at the base, it was intended to be a monument to human ingenuity and a triumph of technology over nature. But as the reservoir behind the dam was about to reach its carrying capacity, leaks began to appear, and on Saturday, June 5, 1976, the dam collapsed, sending a wall of water downstream to wreak death and destruction.

In the 1950s and early 1960s, a young Carl Sagan was like a wild and free-flowing river, a mind gifted with torrents of creativity interspersed with quiet pools of deep reflection. Then, in 1964, a federal agency, the National Aeronautics and Space Administration, without any scientific justification, made the fateful decision to tame that river by building a dam, putting Sagan in shackles that would never be broken. No sooner was it built than leaks began to appear, but NASA, SETI, and professional skeptics looked the other way. They refused to admit that it was a tragic mistake to try to harness a brilliant mind and force it to places it didn't want to be. Without minimizing his incredible accomplishments, which were more than most men or women would dare dream of, the world got a toned-down and scaled-back version of what Sagan could have been had he been allowed the freedom to pursue his dreams.

For the remainder of his life, Carl Sagan was denied his destiny of exploring the possibility of past alien visitations to Earth. Now, in the year 2016, 20 years after Sagan's death, the dam that NASA built is finally about to collapse, and the cascading visions and introspective thoughts of Carl Sagan regarding ancient aliens will, it is hoped, finally be allowed to flow free.

Reprinted from

PLANETARY AND
SPACE SCIENCE

N63 20930

CODE NONE

PERGAMON PRESS
OXFORD · LONDON · NEW YORK · PARIS

Repr. from

Planet. Space Sci. 1963, Vol. 11. pp. 485 to 498. Pergamon Press Ltd. Printed in Northern Ireland

V. 11, 1963 p 485-498

Reprint

DIRECT CONTACT AMONG GALACTIC CIVILIZATIONS
BY RELATIVISTIC INTERSTELLAR SPACEFLIGHT*

CARL SAGAN† *(Harvard O.)*

Department of Genetics, Stanford University Medical Centre,
Palo Alto, California

Stanford U., Calif.

20930

(Received 16 December 1962)

Abstract—An estimate of the number of advanced technical civilizations on planets of other stars depends on our knowledge of the rate of star formation; the frequency of favorably situated planets; the probabilities of the origins of life, of intelligence and of technical civilization; and the lifetimes of technical civilizations. These parameters are poorly known. The estimates of the present paper lead to ~10^6 extant advanced technical civilizations in our Galaxy. The most probable distance to the nearest such community is then several hundred light years.

Interstellar spaceflight at relativistic velocities has several obvious advantages over electromagnetic communication among these civilizations. One striking feature is that with uniform acceleration of 1 g to the midpoint of the journey, and uniform deceleration thereafter, all points in the Galaxy are accessible within the lifetime of a human crew, due to relativistic time dilation. Some of the technical problems in the construction of starships capable of relativistic velocities are discussed. It is concluded that with nuclear staging, fusion reactors, and the Bussard interstellar ramjet, no fundamental energetic problems exist for relativistic interstellar spaceflight.

We assume that there exists in the Galaxy a loosely integrated community of diverse civilizations, cooperating in the exploration and sampling of astronomical objects and their inhabitants. If each such advanced civilization launches one interstellar vehicle per year, the mean time interval between samplings of an average star would be 10^5 years, that between samplings of a planetary system with intelligent life would be 10^4 years, and that between sampling of another advanced civilization would be 10^3 years. It follows that there is the statistical likelihood that Earth was visited by an advanced extraterrestrial civilization at least once during historical times. There are serious difficulties in demonstrating such a contact by ancient writings and iconography alone. Nevertheless, there are legends which might profitably be studied in this context. Bases or other artifacts of interstellar spacefaring civilizations might also exist elsewhere in the solar system. The conclusions of the present paper are clearly provisional.

INTRODUCTION

In recent years there has been a resurgence of interest in the ancient speculation that civilizations exist on other worlds beyond the Earth. This question has retained a basic and widespread appeal from the beginnings of human history; but only in the past decade has it become even slightly tractable to serious scientific investigation. Work on stellar statistics and stellar evolution has suggested that a large fraction of the stars in the sky have planetary systems. Studies of the origin of the solar system and of the origin of the first terrestrial organisms have suggested that life readily arises early in the history of favorably-situated planets. The prospect occurs that life is a pervasive constituent of the universe. By terrestrial analogy it is not unreasonable to expect that, over astronomical timescales, intelligence and technical civilizations will evolve on many life-bearing planets. Under such circumstances the possibility then looms that contact with other galactic communities may somehow be established.

It has been argued that the natural channel for interstellar communication is radio emission near the 21 cm line of neutral hydrogen[1]; or between 3·2 and 8·1 cm[2]; or at

* Based on an address delivered to the American Rocket Society, 17th Annual Meeting Los Angeles, 15 November 1962.

† Department of Astronomy, Harvard University, and Smithsonian Astrophysical Observatory, Cambridge, Massachusetts, U.S.A.

485

10·5 cm[3]. Alternatively, laser modulation of the intensity of core reversal in the Fraunhofer lines of late-type stars has been suggested[4]; or automatic interstellar probe vehicles transmitting a precoded message to planetary sources of monochromatic radio emission which are randomly encountered[5].

The purpose of the present paper is to explore the likelihood and possible consequences of another communications channel: direct physical contact among galactic communities by relativistic interstellar spaceflight. Part of the impetus for publishing these remarks has been a paper by von Hoerner[3] which arrives at very pessimistic estimates for the number of extraterrestrial civilizations; and three papers[6,7,8] which reach distinctly negative conclusions on the ultimate prospect of relativistic interstellar spaceflight. I feel that the information now available permits rather different conclusions to be drawn.

The line of argument to be pursued involves a number of parameters which are only poorly known. The discussion is intended to stimulate further work in a number of disciplines. The reader is invited to adopt a skeptical frame of mind, and to modify the conclusions accordingly. Only through extensive discussion and experiment will the true outlines gradually emerge in this enigmatic but significant subject.

2. DISTRIBUTION OF TECHNICAL CIVILIZATIONS IN THE GALAXY

We desire to compute the number of extant galactic communities which have attained a technical capability substantially in advance of our own. At the present rate of technological progress, we might picture this capability as several hundred years or more beyond our own stage of development. A simple method of computing this number is primarily due to F. D. Drake, and was discussed extensively at a Conference on Intelligent Extraterrestrial Life held at the National Radio Astronomy Observatory in November, 1961, and sponsored by the Space Science Board of the National Academy of Sciences†. While the details differ in several respects, the following discussion is in substantial agreement with the conclusions of the Conference.

The number of extant advanced technical civilizations possessing both the interest and the capability for interstellar communication can be expressed as

$$N = R_* f_p n_e f_l f_i f_c L. \tag{1}$$

R_* is the mean rate of star formation averaged over the lifetime of the Galaxy, f_p is the fraction of stars with planetary systems, n_e is the mean number of planets in each planetary system with environments favorable for the origin of life, f_l is the fraction of such favorable planets on which life does develop, f_i is the fraction of such inhabited planets on which intelligent life with manipulative abilities arises during the lifetime of the local sun, f_c is the fraction of planets populated by intelligent beings on which an advanced technical civilization in the sense previously defined arises during the lifetime of the local sun, and L is the lifetime of the technical civilization. We now proceed to discuss each parameter in turn.

Since stars of solar mass or less have lifetimes on the main sequence comparable to the age of the Galaxy, it is not the present rate of star formation but the mean rate of star formation during the age of the Galaxy which concerns us here. The number of known stars in the Galaxy is $\sim 10^{11}$, most of which have mass equal to or less than the Sun. The age of the Galaxy is $\sim 10^{10}$ years. Consequently, a first estimate for the mean rate of star formation is ~ 10 stars/year. The present rate of star formation is at least an order of

† Attending this meeting were D. W. Atchley, M. Calvin, G. Cocconi, F. D. Drake, S. S. Huang, J. C. Lilly, P. M. Morrison, B. M. Oliver, J. P. T. Pearman, C. Sagan and O. Struve.

magnitude less than this figure, and the rate of star formation in early galactic history is possibly several orders of magnitude more[9]. According to present views of stellar nucleo-genesis[10], stars (and, by implication, planets) formed in the early history of the Galaxy are extremely poor in heavy elements. Technical civilizations developed on such ancient planets would of necessity be extremely different from our own. But in the flurry of early star formation when the Galaxy was young, heavy elements must have been generated rapidly and later generations of stars and planets would have had adequate endowments of high mass number nuclides. These very early systems should be subtracted in our estimate of R_*. On the other hand, a suspicion exists that large numbers of low mass stars may exist to the right of the main sequence in the Hertzsprung-Russell diagram[11]. Inclusion of these objects will tend to increase our estimate of R_*. For present purposes we adopt $R_* \sim 10/\mathrm{yr}$.

There is a discontinuity in stellar rotational velocities near spectral type F5V; stars of later spectral type have very slow equatorial rotation rates. This circumstance is generally attributed to the transfer of angular momentum from the star to a surrounding solar nebula by magnetic coupling[12,13,14,15]. The solar nebula is then expected to condense into a planetary system[15,16,17,18]. The fraction of stars of later type than F5V is greater than 0·98; well over 60 per cent of these are dwarf M stars[19]. It is not known what influence the luminosity of the star has on the subsequent condensation and dissipation of the surrounding solar nebula. We might expect that stars of much earlier type than the Sun readily dissipate their solar nebulae; and that stars of much later type than the Sun dissipate very little of their solar nebulae, thereby forming large numbers of massive planets of the Jovian type. There is good evidence that many of the chemical processes in the early history of the solar system occurred at low temperature[17], and the low luminosity of late type stars is unlikely to impede condensation processes in the solar nebula. We therefore adopt $f_p \sim 1$.

Planets of double and multiple star systems are expected in general to have—over astronomical timescales—such erratic orbits that the evolution of life on them is deemed unlikely[20]. I fail to find this argument entirely convincing; but for conservative reasons it will be included in the discussion. The fraction of stars which are not members of double or multiple systems is $\sim 0·5$[21]. In our own solar system the number of planets which are favorably situated for the origin of life is at least two (Earth and Mars), and the possibility that life arose at some time on the Jovian planets[22] has recently been raised. It is sometimes argued that life cannot develop on planets of M dwarfs, because the luminosity of the local sun is too small. However, especially for Jovian type planets of M dwarfs, the greenhouse effect in a methane-ammonia-water atmosphere should produce quite reasonable temperatures. We adopt $n_e \sim 0·5 \times 2 = 1$.

The most recent work on the origin of life strongly suggests that life arose very rapidly during the early history of the Earth[22-25]. It appears that the production of self-replicating molecular systems is a forced process which is bound to occur because of the physics and chemistry of primitive planetary environments. Such self-replicating systems, situated in a medium filled with replication precursors, satisfy all the requirements for natural selection and biological evolution. Given sufficient time and an environment which is not entirely static, the evolution of complex organisms is apparently inevitable. In our own solar system, the origin of life has probably occurred at least twice. We adopt $f_l \sim 1$.

The question of the evolution of intelligence is a difficult one. This is not a field which lends itself to laboratory experimentation, and the number of intelligent species available for study on Earth is limited. Intelligent hominids have inhabited the Earth for $\leqslant 10^{-3}$ of Earth history.

It is clear that the evolution of intelligence and manipulative ability has resulted from the product of a large number of individually unlikely events. If the history of the Earth were started again, it is highly improbable that the same sequence of events would recur and that intelligence would evolve in the identical manner. On the other hand, the adaptive value of intelligence and manipulative ability is so great—at least until technical civilizations are developed—that, if it is genetically feasible, natural selection is very likely to bring it forth. There is some evidence that surprisingly high levels of intelligence have evolved in the Cetacea[26]. Phylogenetically, these are rather close to hominids; the neuroanatomy of Cetacea brains is remarkably similar to that of the primates, although the most recent common ancestor of the two groups lived more than 10^8 years ago[26]. The Cetacea have very limited manipulative abilities.

Comparison of the rates of stellar and of biological evolution provides some perspective on the probability that intelligence will arise on an otherwise suitable planet. Terrestrial intelligence and civilization have emerged roughly midway in the Sun's residence time on the main sequence. The overwhelming majority of stars in the sky have longer lifetimes than the Sun. With the expectation that the Earth is not extraordinary in its recent evolution but allowing for the fact that apparently only one intelligent phylogenetic order with manipulative abilities has developed, and this only recently, we adopt $f_i \sim 10^{-1}$.

Whether there is one, or several, foci for the line of cultural development which has led to the present technical civilization on Earth is still an open question, depending in part on the extent of cultural diffusion over large distances some five or six thousand years in the past. It appears that little can be gained from speculation on, e.g. whether Aztec civilization would have developed a technical phase had there been no *Conquistadores*.

Recorded history—even in mythological guise—covers $\leqslant 10^{-2}$ of the period in which the Earth has been inhabited by hominids, and $\leqslant 10^{-5}$ of geological time. The same considerations are involved as in the determination of f_i. The development of a technical civilization has high survival value at least up to a point; but in any given case it depends on the concatenation of many improbable events; and it has occurred only recently in terrestrial history. It is unlikely that the Earth is very extraordinary in possessing a technical civilization among planets inhabited by intelligent beings. As before, over stellar evolutionary timescales, we adopt $f_c \sim 10^{-1}$.

The multiplication of the preceding factors gives

$$N = 10 \times 1 \times 1 \times 1 \times 10^{-1} \times 10^{-1} \times L = 10^{-1}L.$$

L is the mean lifetime in years of a technical civilization possessing both the interest and the capability for interstellar communication. For the evaluation of L there is—fortunately for us, but unfortunately for the discussion—not even one known terrestrial example. The present technical civilization on Earth has reached the communicative phase (in the sense of high-gain directional antennas for the reception of extraterrestrial radio signals) only within the last few years. There is a sober possibility that L for Earth will be measured in decades. It is also possible that international political differences will be permanently settled, and that L may be measured in geological time. It is conceivable that, on other worlds, the resolution of national conflicts and the establishment of planetary governments are accomplished before weapons of mass destruction become available. We can imagine two extreme alternatives for the evaluation of L: (a) a technical civilization destroys itself soon after reaching the communicative phase ($L < 10^2$ years); or (b) a technical civilization learns to live with itself soon after reaching the communicative phase. If it survives $> 10^2$ years, it will be unlikely to destroy itself afterwards. In the latter case its lifetime may be measured on a stellar evolutionary timescale ($L \gg 10^8$) years. Such a society will exercise self-selection on its members; genetic changes will be unable to move the species off the adaptive peak of the technical civilization. The technology will certainly be adequate to cope with tectonic and orogenic changes. Even the evolution of the local sun through the red giant and white dwarf evolutionary stages may not pose insuperable problems for the survival of an extremely advanced community.

It seems improbable that, surrounded by large numbers of flourishing and diverse galactic communities, a given planetary civilization will retreat from the communicative phase. This is one reason that L is itself a function of N. Von Hoerner[3] has suggested another reason: he feels that the means of avoiding self-destruction will be among the primary contents of initial interstellar communications.

Gold[27] has talked of the possibility that interstellar space voyagers accidentally may biologically contaminate lifeless planets, and thereby initiate the origin of life. There is also some prospect that such initiation might be purposefully performed. In these cases $f_l = f_l(N)$. Below we will discuss the possibility that $f_c = f_c(N)$. For these reasons it should be remembered that equation (1) is in reality an integral equation.

The two choices for L ($< 10^2$ years and $\gg 10^8$ years) lead to two values of N: < 10 communicative technical civilizations per galaxy, or $\gg 10^7$. Thus the evaluation of N depends quite critically on our expectation for the lifetime of the average advanced community. Von Hoerner[3] has made very pessimistic estimates for L, and his values of N are correspondingly small. It seems more reasonable to me that at least a few per cent of the advanced technical civilizations in the Galaxy do not destroy themselves, nor lose interest in interstellar communication, nor suffer insuperable biological or geological catastrophies, and that their lifetimes, therefore, are measured on stellar evolutionary timescales. Averaged over all technical civilizations, we therefore take $L \sim 10^7$ years. For the purposes of the following discussion then, we adopt as the steady-state number of extant advanced technical civilizations in the Galaxy:

$$N \sim 10^6.$$

Thus, approximately 0·001 per cent of the stars in the sky will have a planet upon which an advanced civilization resides. The most probable distance to the nearest such community is then several hundred light years.*

3. FEASIBILITY OF INTERSTELLAR SPACEFLIGHT

The difficulties of electromagnetic communication over such interstellar distances are serious. A simple query and response to the nearest technical civilization requires periods approaching 1000 years. An extended conversation—or direct communication with a particularly interesting community on the other side of the Galaxy—will occupy much greater time intervals, 10^4 to 10^5 years.

Electromagnetic communication assumes that the choice of signal frequency will be obvious to all communities. But there has been considerable disagreement about interstellar transmission frequency assignment even on our own planet[1,2,3,4,5]; among galactic communities, we can expect much more sizable differences of opinion about what is obvious

* In the Space Science Board Conference previously mentioned, the conclusions for N spanned 10^4–10^9, and the distance to the nearest advanced community ranged from ten to several thousand light years.

and what is not. No matter how ingenious the method, there are certain limitations on the character of the communication effected with an alien civilization by electromagnetic signalling. With billions of years of independent biological and social evolution, the thought processes and habit patterns of any two communities must differ greatly; electromagnetic communication of programmed learning between two such communities would seem to be a very difficult undertaking indeed. The learning is vicarious. Finally, electromagnetic communication does not permit two of the most exciting categories of interstellar contact—namely, contact between an advanced civilization and an intelligent but pretechnical society, and the exchange of artifacts and biological specimens among the various communities.

Interstellar space flight sweeps away these difficulties. It reopens the arena of action for civilizations where local exploration has been completed; it provides access beyond the planetary frontiers.

There are two basic methods of achieving interstellar spaceflight within characteristic human lifetimes. One involves the slowing down of human metabolic activities during very long flight times. In the remainder of the paper, we will discuss relativistic interstellar spaceflight, which, in effect accomplishes the identical function, and further, permits the voyager to return to his home planet in much shorter periods of time, as measured on the home planet.

If relativistic velocities can be achieved, time dilation will permit very long journeys within a human lifetime. Consider a starship capable of uniform acceleration to the midpoint of the journey, and uniform deceleration thereafter. The relativistic equations of motion have been solved by Peschka[28] and by Sänger[29]. For our flight plan, their results are readily modified, and yield for the time t, as measured on the space vehicle, to travel a distance S, with a uniform acceleration a to $S/2$ and a uniform deceleration $-a$ thereafter:

$$t = (2c/a) \text{ arc cosh } (1 + aS/2c^2) \qquad (2)$$

where c is the velocity of light. The results for such an acceleration–deceleration flight plan are shown in Fig. 1.

At an acceleration of 1 g—as would be appropriate for inhabitants of a planet of terrestrial mass and radius—it takes only a few years shiptime to reach the nearest stars, 21 years to reach the Galactic Center, and 28 years to reach the nearest spiral galaxy beyond the Milky Way. With accelerations of 2 or 3 g—as would be appropriate for inhabitants of a planet of Jovian mass and radius—these distances can be negotiated in about half the time. Of course there is no time dilation on the home planet; the elapsed time in years approximately equals the distance of the destination in light years plus twice the time to reach relativistic velocities. For distances beyond about ten light years, the elapsed time on the home planet in years roughly equals the distance of the destination in light years. Thus, for a round trip with a several-year stopover to the nearest stars, the elapsed time on Earth will be a few decades; to Deneb, a few centuries; to the Vela Cloud Complex, a few millenia; to the Galactic Center, a few tens of thousands of years; to M31, a few million years; to the Virgo Cluster of Galaxies, a few tens of millions of years; and to the Coma Cluster, a few hundreds of millions of years. Nevertheless, each of these immense journeys could be performed within the lifetimes of a human crew. For transgalactic and intergalactic distances, equation (2) reduces to

$$t = 2c/a \ln (aS/c^2) \qquad (3)$$

and in this range the curves of Fig. 1 are straight lines on a semi-logarithmic plot.

A number of difficulties have been presented by early authors on the technical aspects of relativistic interstellar spaceflight. Even with complete conversion of mass into energy, extreme mass ratios are required if all the fuel is carried at launch. For relativistic velocities and the above flight plan, the mass ratio is approximately equal to $2/\phi$, where $\phi = 1 - (v/c)$, and v is the maximum vehicle velocity[7,30,31]. For example, to reach $v = 0.999\ c$, the liftoff weight must be some 2000 times the payload, and it is clear that enormous initial vehicle masses are required. For the round trip with no refueling, the mass ratio is $(2/\phi)^2$. Thus, Ackeret[30,31] concluded that "even with daring assumptions," interstellar spaceflight at relativistic velocities would be feasible only for travel to the nearest stars. Furthermore, baryon charge conservation prevents the complete conversion of matter to energy, except

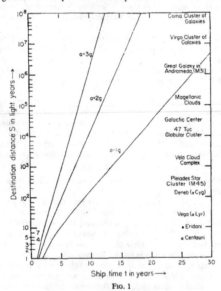

FIG. 1

if half the working fuel is antimatter[32]; the containment of the antimatter—to say nothing of its production in the quantities required—is clearly a very serious problem. An additional difficulty with such an antimatter starship drive has been emphasized by Purcell[7]; the gamma ray exhaust would be lethal for the inhabitants of the launch planet if the drive were turned on near the planet (and if atmospheric absorption is neglected). Staging of fusion rockets[33] provides some relaxation of the required mass ratios, but it appears that relativistic velocities cannot be obtained by such staging alone.

A way out of these difficulties has been provided by Bussard[34] in a most stimulating paper. Bussard describes an interstellar ramjet which uses the interstellar medium both as a working fluid (to provide reaction mass), and as an energy source (by thermonuclear fusion). There is no complete conversion of matter into energy; the existing mass deficits and low reaction cross-sections for the conversion of hydrogen to deuterium are used. The reactor is certainly not available today; but it violates no physical principles, it is currently being very actively pursued, and there is no reason to expect it to be more than a few centuries away from realization on this planet. The Bussard interstellar ramjet requires very large frontal area loading densities: $\sim 10^{-8}$ g cm^{-2} per nucleon cm^{-3} in the interstellar medium. Thus, if the payload is 10^9 gm, the intake area must have a radius of ~ 60 km in regions where the interstellar density is as high as 10^3 nucleons cm^{-3}. In ordinary interstellar space, where the density is ~ 1 nucleon cm^{-3}, the intake area radius must be ~ 2000 km. If the latter radius seems absurdly large, even projecting for the progress of future technology, we can easily imagine the vehicle to seek trajectories through clouds of interstellar material, and vary its acceleration with the density of the medium within which it finds itself. Pierce[6] had earlier considered and rejected interstellar ramjets, but the rejection was based on much smaller intake areas than Bussard proposes.

Of course the intake area may not necessarily be material; to the extent that the ramjet sweeps up ionized interstellar material magnetic fields could be used for collection. Starships would then seek trajectories through H II regions. The Bussard interstellar ramjet also requires moderate liftoff velocities; but even presently-achievable liftoff velocities as low as 1–10 km/sec would be adequate.

Bussard does not discuss the method of funnelling the interstellar matter so it can be collected and utilized for propulsion. Indeed this is one fundamental problem which must be faced by any relativistic interstellar vehicle; otherwise the structural and biological damage from the induced cosmic ray flux will prevent any useful application of the extreme velocities achieved. The maximum velocity of the vehicle in the rest frame, after covering $S/2$, half the distance to the destination, at uniform acceleration a, is given[29,30] by

$$v = c[1 - (1 + aS/2c^2)^{-2}]^{1/2}. \qquad (4)$$

Equation (4) is illustrated in Fig. 2 for the same three choices of a which have already been used. The abscissa gives the maximum velocity reached, expressed as $\phi = 1 - v/c$, during a half acceleration–half deceleration flight plan to a destination at distance S. For example, for a trip to Galactic Center, maximum velocities within 10^{-7} to 10^{-8} per cent of the velocity of light are required. Also shown in Fig. 2 are the velocities at which relative kinetic energies of 1 MeV, 1 BeV, and 1 erg are imparted to interstellar protons by the motion of the vehicle. For travel to even the nearest stars within a human shipboard lifetime, protection from the induced cosmic ray flux is mandatory. It is evident from the large mass ratios already required for boosted interstellar flight, and from the low frontal loading area surface densities required for an interstellar ramjet, that material shielding is probably not a feasible solution.

If some means of ionizing the impacting interstellar material could be found, the ions can be deflected and captured by a magnetic field. In the case that trajectories through H II regions are sought, the interstellar medium will be already largely ionized, and magnetic funnelling would be practicable. The configuration of the field would have to be designed very ingeniously, but the average field strengths required could be as low as a few hundred gauss even for very long voyages. Much higher field strengths would be required, at least in the propulsion module, for a fusion ramjet; or alternatively for a contained plasma driving a photon rocket[25]. It appears likely that superconducting flux pumps[36] can provide the magnetic field strengths required for deflection of the induced cosmic ray flux.

Bussard's[34] concluding remarks on the size of the frontal loading area and the magnitude of the effort involved in relativistic interstellar spaceflight are worth quoting: "This is very large by ordinary standards, but then, on any account, interstellar travel is inherently a rather grand undertaking, certainly many magnitudes broader in scope and likewise more

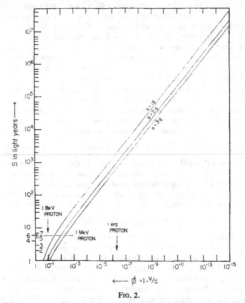

Fig. 2.

difficult than interplanetary travel in the solar system, for example. The engineering effort required for the achievement of successful short-time interstellar flight will likely be as much greater than that involved in interplanetary flight as the latter is more difficult than travel on the surface of the Earth. However, the expansion of man's horizons will be proportionately greater, and nothing worthwhile is ever achieved easily."

The purpose of this Section is to lend credence to the proposition that a combination of staged fusion boosters, large mass-ratios, ramjets working on the interstellar medium and trajectories through H II regions is capable of travel certainly to the nearest stars within a human shipboard lifetime, without appeal to as yet undiscovered principles. Especially allowing for a modicum of scientific and technological progress within the next few centuries, I believe that interstellar spaceflight at relativistic velocities to the farthest reaches of our Galaxy is a feasible objective for humanity. And if this is the case, other civilizations, aeons more advanced than ours, must today be plying the spaces between the stars.

4. FREQUENCY OF CONTACT AMONG GALACTIC COMMUNITIES

We can expect that if interstellar spaceflight is technically feasible—even though an exceedingly expensive and difficult undertaking, from our point of view—it will be developed. Even beyond the exchanges of information and ideas with other intelligent communities, the scientific advantages of interstellar spaceflight stagger the imagination. There are direct astronomical samplings—of stars in all evolutionary stages, of distant planetary systems, of the interstellar medium, of very ancient globular clusters. There are cooperative astronomical ventures, such as the trigonometric parallaxes of extremely distant objects. There is the observation and sampling of a multitude of independent biologies and societies. These are undertakings which could challenge and inspire even a very long-lived civilization.

For the civilization lifetimes, L, previously adopted we see that interstellar space flight to all points within the Galaxy, and even to other galaxies, is possible in principle. The voyagers will return far in the future of their departure, but we have already anticipated that the civilization will be stable over these immense periods of time. There will still be a record of the departure, a repository for the information collected, and a community interested in the results. To avoid unnecessary duplication in interstellar exploration, the communicative societies will pool information and act in concert, as Bracewell[6] has already pointed out. Direct contacts and exchange of information and artifacts will exist among most spacefaring societies possessing relativistic starships. In fact, over large distances, starship communication will occur very nearly as rapidly as, and much more reliably than, communication by electromagnetic radiation. The situation bears some similarity to the post-Renaissance seafaring communities of Europe and their colonies before the development of clipper and steam ships. If relativistic interstellar space flight is feasible, the technical civilizations of the Galaxy will be an intercommunicating whole; but the communication will be sluggish.

It is of some interest to estimate the mean time interval between contacts for a given planetary system. Although the shipboard transit times at relativistic velocities are very roughly the same to any place in the Galaxy, the elapsed time on the home planet is of course approximately proportional to the distance of the voyage. Consequently contact should be greatest among neighboring communities, although we can anticipate that occasional very long journeys will be attempted.

Let each of the N planets in the communicative phase launch q relativistic starships per year. These vehicles each effect at least one contact per journey, and are most often gone some 10^3–10^4 years from the home planet per mission. In the steady state, there are then q contacts effected by each starship-launching civilization per year, and $\sim qN$ contacts per year for the Galaxy as a whole. Relative to the economic capacity of such advanced civilizations, a value of $q = 1$ yr^{-1} seems modest. (Other choices of q will modify the results in an obvious manner.) Each civilization then makes ~ 1 contact per year, and an average of 10^7 contacts during its lifetime. The number of contacts per year for the Galaxy as a whole is then 10^6; a sizable fraction of these should be between two advanced communities. The mean number of starships on patrol from each technical civilization at any given time is $\sim 10^3$–10^4*.

---

* It is easily shown that with the adopted values of N and q, and with even very large ramjet frontal loading areas, the exhaust from such interstellar vehicles makes a negligible contribution to the background galactic cosmic ray flux.

If contacts are made on a purely random basis, each star should be visited about once each 10^5 years. Even the most massive stars will then be examined at least once while they are on the main sequence. Especially with a central galactic information repository, these advanced civilizations should have an excellent idea of which planetary environments are most likely to develop intelligent life. With average contact frequency per planet of 10^{-5} yr^{-1}, the origin and evolution of life on every planet in the Galaxy can be monitored efficiently. The successive development of metazoa, of cooperative behavior, of the use of tools, and of primitive intraspecific communication schemes would each be noted, and would each be followed by an increase in the interstellar sampling frequency. If $f_i \sim 10^{-1}$, then, on a purely random basis, the frequency of contact with intelligent pretechnical planetary communities should be $\sim 10^{-4}$ yr^{-1}. Once technical civilization has been established, and especially after the communicative phase has come into being, the contact frequency should again increase; if $f_c \sim 10^{-1}$, to some 10^{-3} yr^{-1}. Planets of extraordinary interest will be visited even more frequently. Under the preceding assumptions, each communicative technical civilization should be visited by another such civilization about once every thousand years. The survey vehicles of each civilization should return to the home planet at a rate of about one a year, and a sizable fraction of these will have had contact with other communities. The wealth, diversity and brilliance of this commerce, the exchange of goods and information, of arguments and artifacts, of concepts and conflicts, must continuously sharpen the curiosity and enhance the vitality of the participating societies.

The preceding discussion has a curious application to our own planet. On the basis of the assumptions made, some one or two million years ago, with the emergence of *Proconsul* and *Zinjanthropus*, the rate of sampling of our planet should have increased to about once every ten thousand years. At the beginning of the most recent post-glacial epoch, the development of social structure, art, religion, and elementary technical skills should have increased the contact frequency still further. But if the interval between samplings is only several thousand years, there is then a possibility that contact with an extraterrestrial civilization has occurred within historical times*.

5. POSSIBILITY OF EXTRATERRESTRIAL CONTACT WITH EARTH DURING HISTORICAL TIMES

There are no reliable reports of contacts during the last few centuries, when critical scholarship and nonsuperstitious reasoning have been fairly widespread. Any earlier contact story must be encumbered with some degree of fanciful embellishment, due simply to the views prevailing at the time of the contact. The extent to which subsequent variation and embellishment alters the basic fabric of the account varies with time and circumstance. Brailoiu[37] records an incident in Rumanian folklore, where, but forty years after a romantic tragedy, the story became elaborately embellished with mythological material and supernatural beings. At the time as the ballad was being sung and attributed to remote antiquity, the actual heroine was still alive.

Another incident, which is more relevant to the topic at hand, is the native account of the first contact with European civilization by the Tlingit people of the Northeast Coast of North America[38]. The contact occurred in 1786 with an expedition led by the French navigator La Perouse. The Tlingit kept no written records. One century after the contact the verbal narrative of the encounter was related to Emmons by a principal Tlingit chief.

* This possibility has been seriously raised before; for example, by Enrico Fermi, in a now rather well-known dinner table discussion at Los Alamos during the Second World War, when he introduced the problem with the words "Where are they?"

The story is overladen with the mythological framework in which the French sailing vessels were initially interpreted. But what is very striking is that the true nature of the encounter had been faithfully preserved. One blind old warrior had mastered his fears at the time of the encounter, had boarded one of the French ships, and exchanged goods with the Europeans. Despite his blindness, he reasoned that the occupants of the vessels were men. His interpretation led to active trade between the expedition of La Perouse and the Tlingit. The oral tradition contained sufficient information for later reconstruction of the true nature of the encounter, although many of the incidents were disguised in a mythological framework: e.g., the sailing ships were described as immense black birds with white wings.

The encounter between the Tlingit and La Perouse suggests that under certain circumstances a brief contact with an alien civilization will be recorded in a reconstructable manner. The reconstruction will be greatly aided if (1) the account is committed to written record soon after the event, (2) a major change is effected in the contacted society by the encounter, and (3) no attempt is made by the contacting civilization to disguise its exogenous nature.

On the other hand, it is obvious that the reconstruction of a contact with an extraterrestrial civilization is fraught with difficulties. What guise may we expect such a contact myth to wear? A simple account of the apparition of a strange being who performs marvelous works and resides in the heavens is not quite adequate. All peoples have a need to understand their environment, and the attribution of the incompletely understood to nonhuman deities is at least mildly satisfying. When interaction occurs among peoples supporting different deities, it is inevitable that each group will claim extraordinary powers for its god. Residence of the gods in the sky is not even approximately suggestive of extraterrestrial origin. After all, where can the gods reside? Obviously not over in the next county; it would be too easy to disprove their existence by taking a walk. Until very subtle metaphysical constructs are developed—possibly in desperation—the gods can only live beneath the ground, in the waters, or in the sky. And except perhaps for seafaring peoples, the sky offers the widest range of opportunities for theological speculation.

Accordingly, we require more of a legend than the apparition of a strange being who does extraordinary works and lives in the sky. It would certainly add credibility if no obvious supernatural adumbration were attached to the story. A description of the morphology of an intelligent non-human, a clear account of astronomical realities for a primitive people, or a transparent presentation of the purpose of the contact would increase the credibility of the legend.

In the Soviet Union, Agrest[39] and others have called attention to several biblical incidents which they suspect to reflect contact with extraterrestrial civilizations. For example, Agrest considers the incidents related in the apocryphal book, the "Slavonic Enoch," to be in reality an account of the visitation of Earth by extraterrestrial cosmonauts, and the reciprocal visitation of several galactic communities by a rather befuddled inhabitant of Earth. However, the Slavonic Enoch fails to satisfy several of the criteria for a genuine contact myth mentioned above: it has been molded into several different standardized supernatural frameworks; there is no transparent extraterrestrial motivation for the events described; and the astronomy is largely wrong. The interested reader may wish to consult standard versions of the manuscript[40].

There are other legends which more nearly satisfy the foregoing contact criteria, and which deserve serious study in the present context. As one example, we may mention the Babylonian account of the origin of Sumerian civilization by the *Apkallu*, representatives of an advanced, nonhuman and possibly extraterrestrial society[41].

A completely convincing demonstration of past contact with an extraterrestrial civilization may never be provided on textual and iconographic grounds alone. But there are other possible sources of information.

The statistics presented earlier in this paper suggest that the Earth has been visited by various galactic civilizations many times (possibly $\sim 10^4$) during geological time. It is not out of the question that artifacts of these visits still exist, or even that some kind of base is maintained (possibly automatically) within the solar system to provide continuity for successive expeditions. Because of weathering and the possibility of detection and interference by the inhabitants of the Earth, it would be preferable not to erect such a base on the Earth's surface. The Moon seems one reasonable alternative. Forthcoming high resolution photographic reconnaissance of the Moon from space vehicles—particularly of the back side—might bear these possibilities in mind. There are also other locales in the solar system which might prove of interest in this context. Contact with such a base would, of course, provide the most direct check on the conclusions of the present paper.

Otherwise the abundance of advanced civilizations in the Galaxy could be tested by successful detection of intelligible electromagnetic signals of interstellar origin. In the next few decades mankind will have the capability of transmitting electromagnetic signals over distances of several hundreds of light years. The receipt and return of such a signal would announce our presence as a technical civilization, and, if the conclusions of the present paper are valid, would be followed by a special contact mission. Even if an intelligible interstellar signal were received and returned today, it would be several hundreds of years before the contact mission could arrive on Earth. Hopefully, there will then still be a thriving terrestrial civilization to greet the visitors from the far distant stars.

Acknowledgements—This research was supported in part by grant NsG-126-61 from the National Aeronautics and Space Administration while I was at the University of California, Berkeley, where much of the work was performed; the views presented, however, are entirely the responsibility of the author. I am indebted to J. Finkelstein, J. Lederberg, L. Sagan, M. Schmidt, C. Seeger, I. S. Shklovsky, C. Stern and W. Talbert for stimulating and encouraging discussions of various aspects of this paper, and to my fellow members of the Order of the Dolphins, especially F. D. Drake and P. M. Morrison.

REFERENCES

1. G. COCCONI and P. MORRISON, *Nature, Lond.* **184**, 844 (1959).
2. F. D. DRAKE, *Sky & Telesc.* **19**, 140 (1959); and private communication 1961.
3. S. VON HOERNER, *Science* **134**, 1839 (1962).
4. R. N. SCHWARTZ and C. H. TOWNES, *Nature, Lond.* **190**, 205 (1961).
5. R. N. BRACEWELL, *Nature, Lond.* **186**, 670 (1960).
6. J. R. PIERCE, *Proc. Inst. Radio Engrs., N.Y.* **47**, 1053 (1959).
7. E. PURCELL, *Brookhaven Lecture Series* **1**, 1 (1960).
8. S. VON HOERNER, *Science*, **137**, 18 (1962).
9. M. SCHMIDT, *Astrophys. J.*, in press; and private communication (1962).
10. E. M. BURBIDGE, G. R. BURBIDGE, W. A. FOWLER and F. HOYLE, *Rev. Mod. Phys.* **29**, 547 (1957).
11. S. S. HUANG, *Publ. Astr. Soc. Pacif.* **73**, 30 (1961).
12. O. STRUVE, *Stellar Evolution* chap. 2, University Press, Princeton (1950).
13. S. S. HUANG, *Publ. Astr. Soc. Pacif.* **69**, 427 (1957).
14. H. ALFVEN, *Origin of the Solar System*, University Press, Oxford (1954).
15. F. HOYLE, *Quart. J. R. Astr. Soc.* **1**, 28 (1961).
16. G. P. KUIPER, *Astrophysics*, chap. 8 (ed. J. A. Hynek), McGraw-Hill, New York (1951).
17. H. C. UREY, *The Planets*, Yale University Press, New Haven (1951).
18. A. G. W. CAMERON, *Icarus* **1**, 13 (1962).
19. J. H. OORT, *Stellar Populations* p. 415 (ed. D. J. K. O'Connell), North Holland Publishing Co., Amsterdam (1958).
20. S. S. HUANG, *Publ. Astr. Soc. Pacif.* **72**, 106 (1960).
21. C. E. WORLEY, *Astr. J.* **67**, 590 (1962).
22. C. SAGAN, *Radiation Res.* **15**, 174 (1961).

23. A. I. Oparin, A. G. Pasynskii, A. E. Braunshtein and T. E. Pavlovskaya, eds. *The Origin of Life on the Earth*, Pergamon Press, Oxford (1959).
24. J. Oró, *Trans. N.Y. Acad. Sci.*, in press (1962).
25. C. Sagan, C. Ponnamperuma and R. Mariner. To be published.
26. J. C. Lilly, *Man and Dolphin*, Doubleday, New York (1961); and private communication.
27. T. Gold, private communication (1962).
28. W. Peschka, *Astronautica Acta* **2**, 191 (1956).
29. E. Sänger, *Astronautica Acta* **3**, 89 (1957).
30. J. Ackeret, *Helv. Phys. Acta* **19**, 103 (1946).
31. J. Ackeret, *Inter Avia*, **11**, 989 (1956).
32. G. Marx, *Astronautica Acta* **6**, 366 (1960).
33. D. W. Spencer and L. D. Jaffee, *Jet Propulsion Laboratory Technical Rept.* 32–233 (1962).
34. R. W. Bussard, *Astronautica Acta* **6**, 179 (1960).
35. E. Sänger, *J. Brit. Interplan. Soc.* **18**, 273 (1962).
36. S. W. Kash and R. F. Tooper, *Astronautics* **7**, 68 (1962).
37. C. Brailoiu (quoted by M. Eliade) *Cosmos and History*, p. 44 *et seq.*, Harper (1959).
38. G. T. Emmons, *Amer. Anthrop.*, N. S. **13** (1911); reprinted in *Primitive Heritage* (ed. M. Mead and N. Calas) Random House (1953).
39. M. M. Agrest, private communication from I. S. Shklovsky (1962). See also I. S. Shklovsky *Intelligent Life in the Universe*, Holden Day, San Francisco. In press, 1963.
40. W. R. Morfill (translator), *The Book of the Secrets of Enoch* (ed. R. H. Charles), Clarendon Press, Oxford (1896).
41. E. R. Hodges, *Cory's Ancient Fragments*, revised edition, Reeves and Turner, London (1876); P. Schnabel, *Berossos und die Babylonisch-Hellenistische Literatur*, Teubner, Leipzig (1923).

Резюме—Оценка числа передовых технических цивилизаций на планетах других звезд зависит от: нашего знания скорости образования звезд; числа планет в подходящем положении; от вероятностей начала жизни, разума и технической цивилизации и от продолжительности существования технических цивилизаций. Эти параметры мало известны. Оценки данной статьи дают 10^6 еще существующих передовых технических цивилизаций в нашей Галактике. В таком случае наиболее вероятное расстояние от ближайшей общины—несколько сотен световых лет.

Межзвездные космические полеты релятивистскими скоростями обладают некоторыми явными преимуществами перед электромагнитной связью между этими цивилизациями. Одной замечательной особенностью является то, что с постоянным ускорением в силу тяжести до середины путешествия, а после этого с постоянным уменьшением скорости, все точки в Галактике доступны при жизни человеческого экипажа из-за релятивистского расширения времени. Рассматриваются некоторые технические проблемы строительства космических кораблей приспособленных для релятивистских скоростей. Делается вывод, что с ядерными ступенями, термоядерными реакторами и межзвездным прямоточным воздушно-реактивным двигателем Буссарда не существует никаких основных энергетических проблем для релятивистских межзвездых космических полетов.

Мы предполагаем, что в Галактике существует объединенное общество различных цивилизаций, содействующих в исследовании и испытании астрономических объектов и их жителей. Если бы каждая такая передовая цивилизация выпускала один межзвездный корабль в год, то средний временной промежуток между испытанием средней звезды равнялся бы 10^5 годам, между испытанием в планетной системе с различными видами жизни—10^4 годам, а между испытанием другой передовой внеземной цивилизации—10^3 годам. Из этого следует, что есть статистическая вероятность, что какая-то передовая внеземная цивилизация посетила Землю по крайней мере раз в исторические времена. Доказать такой контакт только древними документами и иконографией очень трудно. Тем не менее существуют легенды, которые в этом отношении полезно изучать. Базы или другие искусственные предметы межпланетных цивилизаций могут также существовать где-нибудь и в других местах солнечной системы, напр., на обратной стороне Луны. Выводы настоящей статьи несомненно предварительны.

NOTES

Introduction

1. Sagan, Carl. "Direct Contact Among Galactic Civilizations By Relativistic Spaceflight." NASA archives, 1962.
2. Sagan, Carl, and I.S. Shklovskii. *Intelligent Life in the Universe* (San Francisco, Calif.: Holden-Day, Inc., 1966).
3. Poundstone, William. *Carl Sagan: A Life in the Cosmos* (New York: Henry Holt and Co., 1999).
4. Davidson, Keay. *Carl Sagan: A Life* (New York: John Wiley & Sons, 1999).

Chapter 1

1. Sagan, "Direct Contact."
2. *The American College Dictionary* (New York: Random House).
3. Sagan, "Direct Contact."
4. Ibid.
5. Cory, I. P. "Ancient Fragments, Fragments of Chaldæan history, Berossus: From Alexander Polyhistor." Internet Sacred Archive, *www.sacred-texts.com/cla/af/af02.htm.*
6. Sagan and Shklovskii, *Intelligent Life in the Universe.*
7. Cory, "Ancient Fragments."
8. Kramer, Samuel Noah. *The Sumerians* (Chicago, Ill.: The University of Chicago Press, 1963).
9. The Search for Extraterrestrial Intelligence. NASA Scientific and Technical Information Office, 1977.
10. Drake, Frank, and Dava Sobel. *Is Anyone Out There?* (New York: Delacorte Press, 1992).
11. Poundstone, *Carl Sagan.*
12. Sagan, "Direct Contact."

Chapter 2

1. Drake and Sobel, *Is Anyone Out There?*
2. Sagan, "Direct Contact."
3. Ibid.
4. Drake and Sobel, *Is Anyone Out There?*
5. Ibid.
6. Sagan, "Direct Contact."
7. Drake and Sobel, *Is Anyone Out There?*
8. Ibid.
9. Dyson, Freeman. *The Scientist as Rebel*. New York Review of Books, 2006.
10. Davidson, *Carl Sagan*.
11. Ibid.
12. Ibid.
13. Ibid.
14. Panek, Richard. *The 4% Universe* (New York: Mariner Books, 2011).
15. Drake and Sobel, *Is Anyone Out There?*
16. Sagan, "Direct Contact."

Chapter 3

1. Drake and Sobel, *Is Anyone Out There?*
2. Davidson, *Carl Sagan*.
3. Sagan, Carl. *The Demon-Haunted World* (New York: Random House, 1995).
4. Billings, Lee. *Five Billion Years of Solitude* (New York: Current, 2013).
5. Ibid.
6. Sagan, "Direct Contact."
7. Sagan, *The Demon-Haunted World*.
8. Billings, *Five Billion Years of Solitude*.
9. Borenstein, Seth. "Should We Call the Cosmos Seeking ET?" Associated Press, *The Bend Bulletin*, February 15, 2015.
10. Ibid.
11. Sagan and Shklovskii, *Intelligent Life in the Universe*.
12. Drake, *Is Anyone Out There?*
13. Sagan, Carl. *The Varieties of Scientific Experience* (New York: Penguin Books, 2006).
14. Drake, *Is Anyone Out There?*
15. Michael, George. *Preparing for Contact* (New York: RVP Publishers, 2014).
16. Sagan, "Direct Contact."
17. Drake, *Is Anyone Out There?*
18. Shermer, Michael. *The Moral Arc* (New York: Henry Holt and Company, 2015).
19. Shermer, Michael. *The Believing Brain* (New York: St. Martin's Griffin, 2011).

20. Dawkins, Richard. *The God Delusion* (New York: Houghton Mifflin Co., 2006).
21. Sagan and Shklovskii, *Intelligent Life in the Universe.*
22. Sagan, "Direct Contact."
23. Shermer, Michael. The Moral Arc.
24. Sagan, "Direct Contact."
25. Ibid.
26. Davidson, *Carl Sagan.*
27. Drake, *Is Anyone Out There?*
28. Kurzweil, Ray. Foreword, *The Intelligent Universe,* James Gardner (Wayne, N.J.: New Page Books, 2007).

Chapter 4

1. The SETI Institute Website. *www.seti.org*
2. Sagan, "Direct Contact."
3. Grinspoon, David. *Lonely Planets* (New York: HarperCollins, 2003).
4. NASA Website. *www.nasa.gov.*
5. Davidson, *Carl Sagan.*
6. Ibid.
7. Impey, Chris. *Beyond* (New York: W.W. Norton & Company, 2015).

Chapter 5

1. Sagan, Carl. *The Varieties of Scientific Experience* (New York: Penguin Books, 2006).
2. Ibid.
3. Ibid.
4. Ibid.
5. Ibid.
6. Ibid.
7. Poundstone, *Carl Sagan.*
8. Shermer, Michael. *Why People Believe Weird Things* (New York: W.H. Freeman and Company, 1997), dedication page.
9. Grinspoon, *Lonely Planets.*
10. Shermer, *Why People Believe Weird Things.*
11. White, Chris. "Ancient Alien Evidence Examined," *Skeptic* Vol. 18, No 4 (2013).
12. Shermer, *Why People Believe Weird Things.*
13. Ouellett, Jennifer. Discovery News Website, April 5, 2012. *www.discovery.com.*
14. *Nature* (March 30, 2012).
15. Sagan, *A Life.*
16. Dyson, Freeman. *The Scientist as Rebel* (New York Review of Books, 2006).
17. CSI Website. *www.csicop.org.*
18. Scripp's poll, 2008.
19. NASA Website. *www.nasa.gov.*

20. President Barack Obama press conference, October 19, 2014.

Chapter 6

1. Seldes, George. *The Great Quotations* (N.J.: The Citadel Press, 1983).
2. Davidson, *Carl Sagan*.
3. Jacobsen, Annie. *The Pentagon's Brain* (New York: Little, Brown and Co., 2015).
4. Ibid.
5. Ibid.
6. Davidson, *Carl Sagan*.
7. Jacobsen, *The Pentagon's Brain*.
8. Ibid.
9. Davidson, *Carl Sagan*.
10. Drake and Sobel, *Are We Alone?*
11. Billings, *Five Billion Years of Solitude*.
12. Icarus Interstellar Website. *www.icarusinterstellar.org*.
13. Ibid.
14. 100 Year Starship Project Website. *100yss.org*.
15. Ibid.
16. Ibid.
17. Sagan, Carl. *Cosmos* (New York: Random House, 1980).
18. Davidson, *Carl Sagan*.
19. Basulto, Dominic. *The Washington Post*. As it appeared in the *Bend Bulletin*, December 29, 2015.
20. Feltman, Rachel. *The Washington Post*. As it appeared in the *Bend Bulletin*, January 7, 2016.
21. Impey, *Beyond*.

Chapter 7

1. Sagan, "Direct Contact."
2. Sagan, Carl. Paper. "On the Rarity of Long-Lived, Non-Spacefaring Galactic Civilizations."
3. Sagan and Shklovskii, *Intelligent Life in the Universe*.
4. Ibid.
5. Ibid.
6. *QPB Guide to the Search for Extraterrestrial Intelligence*. Ben Bova and Byron Preiss, editors. Chapter 7 essay, "A Unique Moment in Human History by Michael Michaud" (New York: Simon & Schuster, 1999).
7. Davies, *The Eerie Silence*.
8. See the Templeton Foundation Website. *www.templeton.org*.
9. Michael, George. *Preparing for Contact* (RVP Press, 2014).

INDEX